# DUST

With a Role and a Relationship

Scripture quotations are from the ESV Bible (The Holy Bible, English Standard Version), copyright 2001 by Crossway, a publishing ministry of Good News Publishers. Used by permission. All rights reserved.

Cover art is a digitally altered (by the author) photograph of an anonymous painting in Holland labeled "Vanitas Still Life with Books" 1633.

# DUST
## With a Role and a Relationship:

## Human Identity and Christian Hope in the Face of Death

David A. Norman

2024

# Contents

**Part 1: Getting Started ............... 6**

Chapter 1. Disclaimer: What this Book is For ....... 7

Chapter 2: Biographical Preface ........... 17

Chapter 3: Historical Introduction ........ 29

**Part 2: Perspectives ............... 56**

Chapter 4: The First Person Perspective ........... 57

Chapter 5: The Second Person Perspective ........ 83

Chapter 6: The Third Person Perspective ......... 125

**Part 3: Epilogue ............... 152**

# Part 1: Getting Started

## Chapter 1. Disclaimer: What this Book is For

Have you ever been so cold you struggled to remember what it felt like to be warm? Have you ever been so thirsty you struggled to imagine what it would be like to get to a place where you could fill your bottle with cool, clean water and not feel the need or even desire to drink any of it?

If you have, I hope you didn't stay in that situation very long. It's one thing to be in a bad spot, but when faith starts to lose its grip on hope things can get dangerous.

Spend a few hours in desperate circumstances and we can call it an adventure. Several days and we'd call it hardship. Maybe even trauma. Months? Years? Centuries? How is it that Christians have maintained hope in the face of death with so little evidence for life beyond the grave?

Sure, we might have vague notions of happiness in some sort of new life in the world to come, but how do we get there from here? I mean, I know the "gospel tract" answer. Confess with your mouth and believe in your heart and you will be saved, right? Put your faith in Jesus, and he will give you eternal life. I believe it. I know it. I have been given the gift of faith, and I am extremely thankful for it. I know I will live again after death, but what will that really look like? Who am I, really?

Don't get me wrong. I know myself pretty well. And I have done a lot of thinking about the resurrection and the life to come. But there is a lot about eternal life Christians really don't talk about very much. And I don't just mean the final state. It makes sense that we really don't know a whole lot about that. But when I am faced with death, I don't just want to know where I'll end up. I want to know more about the route from here to there.

I have been trained as a philosopher and a theologian, but sometimes I need more than the ideas. I want something more like a map than a philosophy— and that map may not be perfectly to scale, but it needs a clear legend and a "You are Here" mark so I know to make sense of it.

> When faith starts to lose its grip on hope things can get dangerous.

That is what I will be working toward in this book.

Maybe you are a long distance runner, familiar with the process of pushing your body beyond its natural limits. How do you endure through hardship? When your legs ache and your lungs scream and gravity seems to have been multiplied . . . how do you push on?

If you are like me, it is imagining the future that gets you through. But it's not just any future that gets us through, it's our immediate future. More specifically, what gets me through the enduring grind is imagining the process of how I will get from the agony of exertion to the peace of rest on the other side. Not just the end state, but what I have to get through to make it there.

There is this hill . . . and then one more. Then that long flat bit, that'll be OK. Then around the slow curve . . . then just that awkward stumble/sprint to the finish line and I can breathe and walk and then lie down. Then I'll get a shower and feel great about this day for weeks and months and years to come.

If I don't know the route that well, I may play other mental tricks. Abstraction, for example. Just 10 more minutes. That's just the time it takes to get from my house to pick up my son at school. So I'm getting in the car. I'm backing out the driveway. I'm turning up the hill. Just keep pushing and I will be there in no time. I'll be able to see the finish line. Then I'll cross it. It will hurt like crazy, but it will be a good hurt, because I will be done. I will have met my goal. Then I can eat at least half of that huge, awesome pizza! I can look back on all of this and tell (with joy and happiness) stories about how bad it hurt.

Or if abstraction isn't working, I can straight-up lie to myself. "I'm not really here, running down this trail. This doesn't really hurt. I am in a happy place. I am lying on a beach, under an umbrella. The wind is blowing softly and the warm sun is shining brightly. I just have to relax and keep breathing . . ."

Don't get me wrong, hope is always a good thing. I love hope. Especially when I am suffering. But these three strategies for maintaining hope in the face of suffering are neither morally nor functionally equal. Knowing and clinging to the truth is both morally and functionally better than the best abstraction induced distraction. And it is far superior– both morally and functionally– to even the most creatively constructed and best intentioned lie.

Hope isn't a strategy. At least not a very good one.

For my faith to keep its grip on hope, I need to know the process I'll go through to get from here to there. That's true of running. It's true of being cold. Of being thirsty. Even of facing death.

That's what this book is for.

In the pages that follow, I will try to get as clear as possible about what the Bible teaches about death and the world to come. Not just the source of our salvation, though we'll touch on that. And not just what life beyond the grave will look like, though we'll have a few things to say about that as well. But what I am most interested in is the question of what a human being really is and what happens to us when we die. Getting at these questions is what this book is for.

\* \* \* \* \* \* \*

The central thesis of this book, that we are dust with a role and a relationship, is both ancient and novel.

As we will soon see, it is based on implicit wisdom that is as old as human grammar itself and I believe it is more consistent with the teachings of the Christian Scriptures than any other related thesis I have ever considered. Getting my head around it and working out the details hasn't been easy, but I've taken my time and worked things out as clearly and simply as possible.

In the pages that follow, I will do all I can to keep things clear, simple and light-hearted. In particular, I have tried to keep my references to other writer's work to a very bare minimum. I didn't want this to become a boring, technical work of scholarship. That's not what this book is for.

In this work, I'm not out to prove anyone wrong or even myself right, for that matter. I am just trying to offer some clear solutions to some pretty darn important and difficult problems. That should be enough.

Don't get me wrong– a good, tough, spirited debate can be a lot of fun. But . . . that's not the kind of fun I want to have here. That's not what this book is for.

So what is it for? I want to offer a fresh perspective on some ideas that I think need some re-thinking.

One thing you might as well get used to, therefore, is taking a familiar-sounding word and giving its meaning a nice, long stretch. My hope is that in doing so, our understanding can stretch and grow as well.

Sometimes this stretching will really just be playful– like when, in chapter 3, I refer to the 1,000 years of "Christendom" as "the millennium". I don't literally mean that the 1,000 years between Constantine and the Reformation are the years referred to in Revelation 20:2-3. Maybe they really are the same, but it is more likely that they just rhyme. See what I mean?

Other times, I will be using familiar words in ways that seem (and probably are) slightly unfamiliar– like when using the terms 'mind' and 'soul'. In those cases, I will be as clear as possible regarding my intention to refine the way these words should be used.

Astute readers may have noticed that I already started stretching the meanings of worlds when I said I didn't want this book to be a piece of scholarship. "Of course it's scholarship," you might be saying. "Death and human identity are super-heavy topics. Any book that attempts to address them is going to be a heady piece of work. And it needs to address the thousands of years of work that has already been done on the subject."

> *The central thesis of this book, that we are dust with a role and a relationship, is both ancient and novel.*

If that's what you are thinking, this disclaimer is for you. If you believe we can only adequately address issues of human identity and death with a serious, somber, scholarly tone and tons and tons of nuances, qualifications, proofs and footnotes . . . well . . . sorry. That's not the plan.

So what is the plan? Well . . . while I am stretching the meanings of words . . . here is a real zinger: This book will be more like prophecy than scholarship.

Ahhh! There's an old fashioned word for you! "Prophecy"!

So wait . . . what? How is this supposed to be a book of prophecy? Am I claiming I have been given words from God to inform you regarding what to expect in the future?

Well . . . yes, actually. But not exactly how you might expect.

The words from God I will be relying on have not just been given to me. Far from it. In fact, they form the book that is literally the best selling book of all time. We typically refer to it as "The Book", "The Bible", and sometimes even "The Holy Bible".

I will be relying on other historical books as well, but I'll do all I can to stay WAY out of the weeds. And the only book I will be treating as any sort of authority is the Bible.

Also, the future I will be talking about isn't some major world event like the ones we typically think of when we hear about prophecy. The future we'll be talking about is each of our futures. Our deaths. And our resurrections.

I don't know *how* any of us will die. But that's OK. In this context, "how" isn't nearly as interesting to me as "that"– and especially what that will do to our personal identities.

But I've also got something else in mind. When I say that I want this book to be more like prophecy than scholarship, I am not just saying that I will be relying on God's words to discuss the future.

I am trying to offer a disclaimer. In particular, I want to admit right up front that, while I will be trying very hard to change the way you think about some pretty darn important stuff, I won't be offering you very much in the way of what might traditionally be referred to as evidence. Sorry about that, but . . . evidence is boring and— particularly regarding subjects of as much import as this— tremendously overrated. And I guess I just sort of have to lay that out there as a stand-alone claim because if I could prove it, the proof would be self-defeating.

See, here's the thing: scholarly books are usually filled with all sorts of references to the works of other scholars. Studies, authorities, surveys, discoveries, experiments and lots of other boring stuff. And every page is filled with citations.

This won't be like that.

I will try to explain some pretty complex stuff as simply and clearly as possible and you can either agree (and be informed) or disagree (and explain the stuff better– in which case you will also be informed, and might drop me a note to return the favor). But let's try to keep it friendly, OK? The subject is dark and scary enough without us bringing additional gravity to the scene.

Even more importantly, especially as it relates to this disclaimer, I've gotta confess, I've got some baggage. In particular, I have had conversations about this stuff with lots of people in the past who have gotten really upset with me. I really hope you, dear reader, will not be added to that company. I don't want to be controversial. OK, maybe a little controversial. But I definitely don't want to be pretentious or rude or . . . get you overly upset with me. I want to be helpful. I want to be friendly.

And I definitely don't just want to add to a very long, boring, scholarly discussion. I want to answer important questions. And if my answers aren't right, I want to be forgiven and invited to try again.

This doesn't have to be all scary and careful and serious!

After all, as we will see over and over in the pages that follow, the promise of forgiveness and the sure hope of resurrection make all sorts of things that would ordinarily be paralyzingly scary much, much less so.

## Chapter 2: Biographical Preface

My grandfathers were both technicians. My father and father-in-law are both engineers.

I was raised to be a problem solver.

Some problems are complex. Some are simple. Some problems are trivial. Some are definitely not trivial. I have spent most of my life focused on two related problems that are neither simple nor trivial.

    1. Our existence is clearer than our essence.

    2. Our desire for survival exceeds our understanding of our own persistence conditions.

We can simplify these problems a bit by thinking of them in the form of these questions:

    1. What does it mean to be human?

    2. Is there anything beyond death?

These questions may be familiar to you. They are often asked but rarely answered. This book will be an attempt to answer them clearly, frankly and soberly. If you think these questions are important, I hope you will not be disappointed. I have been working on this stuff for the last 30 years. This book is my attempt to share what I have learned so far . . . and to start a conversation that I believe the Christian church really needs to have right now.

\* \* \* \* \* \* \*

My earliest memories are filled with theological questions. I didn't necessarily think of them as theological at the time, but they certainly were. My parents were part of the Charismatic movement of the 1970's, so some of the earliest questions I remember being batted around were, "Was that God speaking, or some sort of mental illness?", "Was that a demon? And if so, how do we fight it?", "Is that guy really anointed to lead us, or does he just need a job?".

Some questions had simple, practical answers. Santa? The Easter Bunny? The Tooth Fairy? Not worth it. I was told not to believe in them, lest I later think God belonged in that same group of characters. Halloween and trick-or-treating? Definitely not worth it. Not even worth discussing. Rock music? Civic organizations? Theological training? Questionable. Maybe OK, but probably just for liberals who were too rich to have better things to do.

> *But sometimes what God says is complicated and hard to understand. In those situations, it is hard to know how to believe well.*

My answers to some of those "simple" questions have changed over the years but, honestly, probably not as much as you might think. I'm still not a huge "secular holiday" fan, for example, but I have learned to pick my battles . . .

The core theology I grew up with (and still cling to today) is both practical and simple. What God says is true: basically by definition. But which god is the true God? I believe that the god of the Bible is the true God. So practically speaking, when trying to sort out questions of faith and practice, the bible is our only absolute standard of truth.

God said it. I believe it. That settles it.

But sometimes what God says is complicated and hard to understand. In those situations, it is hard to know how to believe well. What we believe gets confusing and things become troublesomely unsettled. Especially in those times, we have to think and study hard. We have to be brutally honest with ourselves even as we remember that grace and humility can give us the patience to persevere.

\*    \*    \*    \*    \*    \*    \*

My grandfather, who grew up in a North Georgia mill village, was orphaned in the 6th grade. He lived "here and there" until he was able to join the Navy at age 17 to fight in World War II.
He was a practical man who grew philosophical later in life– about the time I entered high school.

My grandfather did well for himself. He was trained to be a communications expert in the Navy, working on radios and other, more complex communications and navigation equipment. After the war he stayed in the service training others.  He also began teaching and investing in real estate on the side.

By the time I came to be aware of such things, he was a wealthy man.  He bought a new Cadillac every year, took my Gramma on trips to the Holy Land and other exotic places, and owned a remote cabin on a rural lake that he let us use as often as we wanted.  When I was in High School, he bought an electric organ and started learning to play it.  He grew deeply interested in the power of positive thinking and grew more and more generous.  "This world, this body, and this business" he would say, "are all passing away.  The spiritual world is all that really matters."

Then he started getting . . . even more generous.  My uncle discovered he had been practically giving away houses to people he thought needed a break.  Then he started putting up signs in the house, reading "North", "South", "East" and "West" to keep him from getting disoriented.  He started crashing his Cadillacs– sometimes committing hit-and-run crimes without realizing what he had done.  He got more and more confused, more and more frustrated.  We had to put him in an institution and wait for him to slowly die.

> *If we are confused about what a human being is, we won't know what to hope for when we see them break.*

It turns out, this world and these physical bodies matter a lot more than he had been thinking.

My grandfather was a great man who died a horrible, confused and confusing death. I saw that (and still see that) as a real problem. If we are confused about what a human being is, we won't know what to hope for when we see them break. We will talk a lot about that in this book.

\*   \*   \*   \*   \*   \*   \*

At the risk of sounding a bit . . . dark . . . have you ever seen anyone die? Have you ever touched a dead human body or at least seen one up close? If you have, you are probably a bit triggered now. Sorry about that, but this is important. If you haven't, you are at a significant disadvantage when it comes to understanding the visceral reality of identity and persistence questions.

You may believe in thestrals. You might even understand the practical role they play in our economic life. But if you can't see them for yourself just yet, you are both blessed and handicapped at the same time. There are some aspects of reality you will have to accept on the testimony of others.

I have told you about my paternal grandfather's slow and confusing death. My maternal grandfather died much more quickly. I was only five or six years old, so most of my memories of him are fuzzy. I remember the jar of lemon drops he kept by his recliner and how generous he was with them. I remember him having nose bleeds that wouldn't stop because of his cancer treatments. Then I remember seeing him in an open coffin.

My mother was with me when it was our turn to say goodbye to him. She was crying because she missed him already. And she was crying a lot.

But he was lying right there in front of us– there at the front of the church.

She sobbed and sobbed. I rested my hands on the edge of the coffin as I looked at him. His color was a little funny– pale and maybe a little bit green– but it was definitely him. Other than being in a new gray suit and a little something funny going on with his lips, he just looked like he was sleeping there while the grownups were all talking about him. I probably knew I shouldn't have, but I couldn't help it. I reached out and touched his chin.

I still remember the shock of that touch today. Vividly. Viscerally. PawPaw wasn't sleeping. His skin was cold and felt . . . sort of waxy. The word that jumped in my head was "fake". Somehow, he felt fake.

Was that really him? *Something* was definitely fake, but what? Was this thing not real? This . . . corpse . . . that I had just touched? It was definitely real, but . . . what was so undeniably "fake" here? Where was the real PawPaw? Had he turned into this? Who/what was he?

He was dead, but this was definitely not what I had been taught to expect death to be like. So much about this whole situation was deeply unsettling. I didn't even know how to think about what I had learned. So I did the same thing grownups do. I buried the questions I didn't even really know how to ask and got about the business of life.

Maybe you're thinking this whole project is just my attempt to cope with childhood trauma. Maybe there is something to that. But death is hardly abnormal. Why are we so afraid to talk about it? Is it natural? Is it inevitable? Is it permanent? Do we know anything for sure about personal existence beyond death?

If you haven't already, take a break from reading for a minute and reflect on your own experiences with

> *I buried the questions I didn't even really know how to ask and got about the business of life.*

death. We are going to be exploring answers in this book. But before we can fully apprehend the truth of an answer, we need to experience the question. What is a human being? And what exactly does death do to the personhood of its victim?

\*   \*   \*   \*   \*   \*   \*

When I went to college, I met professors who had devoted their lives to the question of what a mind is and how it relates to a brain. I was hooked. The philosophically "established" position on these questions is known as "non-reductive physicalism". Put simply, this position states that human minds are always dependent on human brains, though talk about minds cannot be reduced to talk about brains.

Books and books and papers and papers have been written, explaining why this position is really the only position that makes sense of the evidence. By the mid-1990's, when I was in college, there was about as much evidence for non-reductive physicalism as there was for heliocentrism in the 1600's. The problem for me was that none of these "non-reductive physicalist" philosophers were evangelical Christians. The evangelicals were still holding on to an older philosophy, known as "substance dualism", because they believed (not entirely unlike my paternal grandfather) that minds and souls were more or less the same thing– so our hope for eternal life in the face of bodily death consists in our mind/souls being able to live on without our dead brains.

I thought my fellow evangelicals were confused. I thought that confusion would lead to more horrible, confusing deaths like my grandfather's. I believed this was very, very important.

But what was I to do about it?

The first step was to clarify my own thinking. I worked hard in college, graduating summa cum laude with a degree in Philosophy. Then I went to seminary, then on for a PhD. I wrote and published a few philosophical and theological papers and a big, boring academic book. (If this book isn't scholarly enough for you, you might like that one better. It is over 300 pages, has 430 references and 548 footnotes.)

But even though I was keeping them at an academic level, my ideas were more controversial among Christians than I wanted them to be. I wanted to bring peace and clarity, not discord and confusion.

So I decided to be patient. I waited 20 years.

Well, I wasn't just *waiting* that whole time, exactly. I have been a Philosophy professor, an executive director, an entrepreneur, president of a 180 year old college and seminary, a middle and high school principal, a full time fundraiser, a Sunday school teacher, an absolutely smitten husband and a proud dad of three incredible young men. But after 20 years, I believe the time has finally come to write this book.

> *I thought my fellow evangelicals were confused.*

Philosophically speaking, this book will be an explanation and defense of the version of non-reductive physicalism I have developed over the years. More importantly (to me, at least), it will come from an openly evangelical point of view.

But I hope this isn't primarily an academic book about philosophy. My aims are much higher than that. I hope this book will become the book I wished I had when my paternal grandfather was dying. It will address the questions that have burned within me ever sense. Most importantly, if the Holy Spirit uses this work as I hope, it will radically affect (and clarify) the way we think about death.

If you are facing the death of a loved one right now, I am deeply sorry. We can try to dress it up and make the best of it all day long, but the simple, undeniable truth is that death is horrible. It is our enemy. We should hate it. I hate it. God hates it. But my wildest hope for this discussion is that it will open a path for us from fear, confusion and desperation in the face of death to confidence, clarity and a more tangible understanding of the Gospel than we ever thought possible.

Here is how our discussion will go:

In the first chapter, I will offer a brief summary of human history through the late 1600's. We will address such questions as, "What can we learn about human identity from the Bible's account of creation?," and "Where did evil come from, and how does that affect our understanding of human experience?"

> But my wildest hope for this discussion is that it will open a path for us from fear, confusion and desperation in the face of death to confidence, clarity and a more tangible understanding of the Gospel than we ever thought possible.

Just in case those questions seem too simple, we will also talk about the rise and fall of centralized church authority and how Modernism grew from the shambles of that authority.

As we discuss the rise of Modernism, we will highlight the central role played by the first person perspective. This should set us up for what I think is the clearest way to understand the relationships we often worry about when it comes to human identity– the relationships between mind, soul and body.

Across cultures and throughout time, human language has generally recognized three linguistic perspectives representing the personal relationship between the speaker and the subject being discussed. While there are always exceptions and qualifications, those perspectives are generally classed as the first, second and third person perspectives. I think we should understand ourselves that way too.

This approach may sound pretty confusing at first, but we will have quite a bit more to say about it as we move forward. Hopefully, it will begin to make a lot more sense as we go. In fact, we will spend three chapters discussing it.

> *Across cultures and throughout time, human language has generally recognized three linguistic perspectives representing the personal relationship between the speaker and the subject being discussed.*

At the risk of sounding like we are behind already, Chapter 4 will discuss the first person perspective. Chapter 5 will discuss the second person

27

perspective.  And Chapter 6 will discuss (you guessed it) the third.  See, you are catching on already!  Finally, by way of conclusion, we will discuss a few very practical matters of particular concern.

My goal is to obviate confusion around the questions I believe are most important, so I promise to be as clear and concise as possible.  I'm not a particularly morbid guy, and I definitely don't want to stir up contention, but there may be aspects of my thinking that are still controversial.  You may find yourself disagreeing with me.  That's OK!  That just means at least one of us doesn't have the full problem solved yet.  But I am sick of all this just being up in my head (with the early ideas stuck on the back shelves of a few research libraries).  Writing this book will be hard work.  Reading it won't be easy either!  Thank you for taking the time.

## Chapter 3: Historical Introduction

From around 500 BC to a little before 2000 AD, the best philosophers in the Western tradition taught that the most fundamental dualism in the universe was between "the physical" and "the spiritual". Christian theology leaned on that broad philosophical consensus with more or less conviction from time to time throughout that broad span, but . . . physical/spiritual dualism is not what the bible actually teaches. The Bible teaches that the fundamental dualism in the universe is between creator and creature.

## 1. Creation

According to the bible, human beings were carved out of the rest of creation to bear the image of the creator. We are dust, with a very important role to fill, for which we are empowered by a very special relationship.

Understanding beginnings is always difficult. For a creature, understanding the beginning of creation is especially difficult. The bible just says, "In the beginning, God created the heavens and the earth." But who is God? And . . . the beginning . . . of what? The earth and its atmosphere? The solar system? The universe? All time? . . . Is that even a thing?

As we keep reading the bible, we eventually pick up on the idea that God is both eternal and triune. Lots of great work has been done on the nature of

God, so we could spend tons of time talking about it. But we won't. That's not really our focus here. Within the community that treats the Bible as sacred, the eternality and trinity of God are uncontroversial– so we will just take this much as a given for our purposes.

So the eternal, triune God was there in the beginning. At least that should help us with our questions about time. After all, time is just a very specific way of talking about relationships. It is not an object (like a pen or a pen cap); it is a concept (like the fit between the pen and the cap). Saying it this way may sound a little strange, but I am not saying anything that should be terribly controversial. Speed is a closely related concept that works the same way. When we talk about speed, we understand readily enough that we are talking about the relationship between one object (or set of objects) and another object (or set of objects). Speed isn't an object in and of itself. Neither is time. They are just concepts.

> *The Bible teaches that the fundamental dualism in the universe is between creator and creature.*

Because God is both eternal and triune, time is eternal. While time is just a concept, it is interesting to note that time does take on some of the attributes of its subjects. When talking about creation, for example, which has a maximum speed, time has a maximum speed. Modern Physics has used this principle to extrapolate some pretty interesting implications. But when we are talking about time before creation, most of those familiar attributes lose their ground (sorry about the pun there!). There is not much we can say. The triune nature of God tells us that there was more than one person before creation, so it follows that there has always been such a thing as time, but . . . We simply don't know much about it. By definition, no part of creation has experienced time before creation, and God hasn't told us about it– presumably because we wouldn't understand Him if He did.

So for our purposes . . . here is the point: even though time was around before creation, it makes sense to simply refer to the initial point of creation as "the beginning". Creation hasn't always been around. There was a beginning. And in the beginning, God created the heavens and the earth. Nothing (no created thing, that is) was before that– just God and a bunch of concepts. This idea is super important to Christian theology, and has been established long enough that the shorthand we normally use to refer to it is in latin– "creation ex nihilo".

> *The heavens and the earth were created out of nothing. Human beings weren't created like that.*

God created the heavens and the earth out of nothing. Then, later, he created human beings. How much later? Well, after light, the sky, land and sea, plants, the sun and stars, birds and fish, animals . . . the main thing to notice here is that it was later. The heavens and the earth were created out of nothing. Human beings weren't created like that. The bible is very clear about human beings being created out of something– out of stuff that had already been created. And the first human beings were named Adam and Eve.

Human beings were formed out of the rest of creation, so as far as objects go . . . I am one of them. And so are you. We are clay. We are ashes. We are dust. This is the way the bible talks about human beings. It would take us too far off task to work through the details here, but a look at the Hebrew for these terms would show us that they are all pretty closely related– the point being that we are not merely residing in stuff, we ARE stuff. We are dust, and to dust we will return.

But how and why were we formed out of the rest of creation? Well, the bible tells us why first, so let's start there.

Genesis 1:26 says, "Then God said, 'Let us make human beings in our image, after our likeness. And let them have dominion over the fish of the sea and over the birds of the heavens and over the livestock and over all the earth and over every creeping thing that creeps on the earth." Put in more modern cultural terms, what is being said here is that God created human beings to be his representatives amid his creation. As becomes even more clear in Genesis 2, we were made to tend to creation– to fill it and to subdue it with the love and the justice of God, not in our own image (trying to make creation more like we want it to be) but in God's image (trying to make creation more like he wants it to be).

So we were formed to play a certain role. But how? How were we formed, and how are we to fulfill the lofty role we were created to fill? Actually, the answers to both of these 'how' questions are intertwined: We were formed by God with special intimacy.

As we just discussed, theologians often talk about the importance of "creation ex nihilo", underscoring that God alone is eternal. Another term that is very important to our work is "the special creation of Adam and Eve". It is out of the specialness of this special creation that human beings are empowered to carry out our purpose as divine image bearers.

Genesis 2:7 says, "then the Lord God formed the man of dust from the ground and breathed into his nostrils the breath of life, and the man became a living creature."

As we mentioned above, the term "dust of the ground" here just means "the stuff of creation". The Bible isn't saying that Adam was terracotta, necessarily, it is just saying that (if you'll forgive the double negative), he didn't spring forth out of nowhere. When we hear about "the breath of life", something similar is going on. Just like the Biblical words for ashes, dust, clay and earth are related (usually just meaning the "stuff" of creation) the concepts of breath, wind and spirit are intertwined in the Bible, and they carry with them a sense of life, action or empowerment.

So when the Bible says, "God breathed into his nostrils the breath of life" it means that this specific bit of creation was given life. But it wasn't given life in exactly the same way "the birds of the air and the fish of the sea and every creeping thing that creeps on the earth" were given life. When it comes to the divine image bearer, we are dealing with a piece of creation that is specially set aside and specially enlivened– directly by the creator.

> We were formed by God with special intimacy.

As I mentioned at the beginning of this chapter, most philosophers from around 500 BC to a little before 2000 AD believed that the universe consisted of two kinds of substance– matter and soul, or the physical and the non-physical. This way of thinking is grounded in the Greek philosophy generally attributed to Plato, who lived around 350 BC – before the Judeo-Christian intellectual tradition had much impact– and it has brought a lot of confusion to the history of Christian theological anthropology.

When most Christian theologians over the last 1,500 years have talked about the special creation of Adam, they usually brought this Greek understanding of the universe to the text– assuming that God breathing "the breath of life" into Adam was God creating a non-physical "thing", or "soul", and implanting it into the clay to make human beings a compound of two substances. But if you don't bring the philosophical presupposition of body/soul dualism to the text, you will see that that is not what the bible is actually saying at all. The Bible doesn't imply that a soul is a different kind of substance. A soul is simply a life. The Bible says that Adam was separated from the rest of creation and given life directly by God.

God made the stuff of creation with at least some form. Formed stuff we fairly loosely call "things" or "objects". Some objects have particularly complicated and active forms. These objects we call "living beings". The ancient Greek philosopher Aristotle talked a lot about these living beings. He referred to their forms (the particularly complicated and active forms of living beings) as "souls". Nowadays,

> *We aren't ghosts trapped in stuff, we are stuff that's been given a special calling and special empowerment.*

we usually just refer to these as "lives", reserving talk of "souls" for humans. After all, the bible demonstrates that some living beings are special. They have forms that have been shaped (ie, made particularly complicated) and activated in an especially intimate way by the creator. These living beings we call divine image bearers . . . or just humans. Human lives are special lives, so we usually reserve the word "souls" for them.

So, to wrap all this up, as far as objects are concerned, we are simply dust. But we have a very big role to fill, and we are especially empowered by the creator himself to do so. That's what human souls are all about. We *aren't* ghosts trapped in stuff, we *are* stuff that's been given a special calling and special empowerment. Put simply, we are dust with a role and a relationship.

And one more thing: God didn't just create a man to bear his image, he created a community. He's not a single person, and neither is his image bearer. He created his image bearers male and female: Adam, Eve . . . and their progeny. This community was intended to grow. And it did. But by the time Moses (who initially authored the first few books of the Bible) came on the scene some painful lessons had come their way.

"But wait," some may say. "You are making it sound like humans are just like animals who die and are no more. But what about the immortality of the human soul?"

I can assure you that many, many very smart people believe in the immortality of the human soul! If you are one of them, please do not take offense at my disagreeing with you. It is only very recently in the history of philosophy that the consensus has shifted from those who, following the ancient Greeks, believe in the natural immortality of the soul to those who don't. I don't mean to imply that I want to simply follow the philosophical trends here, I am just wanting to be as clear as possible about what the Bible actually teaches. If you survey the Bible for talk about immortality (especially the

> God alone is immortal, and He lives in unapproachable light.

Greek philosophical idea that humans have eternal or even simply immortal souls) you won't just find that such talk is not there, you will find the idea flatly contradicted.

We will talk a lot more about this in future chapters, but for now let me acknowledge that the New Testament does teach that human immortality will come at some point in the future. In the Garden of Eden, which we will discuss in more detail shortly, there were two trees at the center: the tree of the knowledge of good and evil and the tree of life. Adam and Eve ate from the tree they were commanded to avoid, but they didn't eat from the tree of life. That is why we are mortal now. In Genesis 3:22-24 we read:

> Then the LORD God said, 'Behold, the man has become like one of us in knowing good and evil. Now, lest he reach out his hand and take also the tree of life and eat, and live forever—' therefore the LORD God sent him out of the garden of Eden to work the ground from which he was taken. He drove out the man, and at the east of the garden of Eden he placed the cherubim and a flaming sword that turned every way to guard the way to the tree of life.

But the tree of life isn't just mentioned in Genesis. It shows up again at the end of the Bible, in Revelation. Jesus, with his life, death and resurrection, reopened the path to eternal life. When he returns, and the heavens and the earth are united, the tree of life will be at the center of the New Jerusalem. It will be amazing! But for now, perhaps the clearest biblical teaching on human immortality comes from 1 Timothy 6:10ff, which says:

> For the love of money is a root of all kinds of evil. Some people, eager for money, have wandered from the faith and pierced themselves with many griefs. But you, man of God, flee from all this, and pursue righteousness, godliness, faith, love, endurance and gentleness. Fight the good fight of the faith. Take hold of the eternal life to which you were called when you made your good confession in the presence of many witnesses. In the sight of God, who gives life to everything, and of Jesus Christ, who while testifying before Pontius Pilate made the good confession, I charge you to keep this command without spot or blame until the appearing of our Lord Jesus Christ, which God will bring about in his own time– God, the blessed and only Ruler, the King of kings and Lord of lords, who alone is immortal and who lives in unapproachable light, whom no one has seen or can see. To him be honor and might forever. Amen.

We will take hold of eternal life eventually, but that's not who we naturally are. God alone is immortal, and He lives in unapproachable light.

Before we spend too much time thinking about the end of this present age, I want to get a few more key points from history on the table. In the next few sections, we will quickly review somewhere between four and ten thousand years worth of history. We will move pretty quickly, but hit all the "high points" essential for evangelical Christianity.

I may tell some parts of the story differently than the ways you've heard it before. Sometimes that will be intentional, but sometimes it will just be accidental. You may be tempted to skim over details, thinking that what follows is just an unnecessary review. I certainly can't stop you from doing that, but please keep this in mind: the main reason I want to review this big story is to prove a point. The story has been told many, many times before with the underlying assumption that human beings are essentially non-physical things temporarily inhabiting bits of dust. I want to cover the main points of sacred history without lapsing into that kind of dualism. Doing so should form a healthy review for all of us while simultaneously demonstrating that the whole metanarrative makes sense without deviating from our central anthropological thesis: that human beings are essentially dust with a role and and a relationship, not composites consisting of a non-physical thing (a mind or soul) temporarily inhabiting a physical thing (a body).

Let's dive in.

## 2. The problem of evil

When Moses sat down to write the Bible, he was in a tough spot.

Despite his race, he had been raised with extreme privilege in Pharaoh's household. But then he'd seen people from his race brutalized. So he murdered one of the abusers. Then he ran away before he could get caught and started a new life for himself. But then he was called by God to return to Egypt and lead all of his race to a new life in a new land.

God called Moses for a special purpose and empowered him with the inspiration of the Holy Spirit to fulfill that purpose. Moses needed to convince his people that leaving Egypt and possessing Canaan was God's plan for Israel. So he sat down, pulled out a pen and some papyrus, and got to work (being filled with the Holy Spirit so powerfully that he was kept from any written error and able to write things he could not possibly have known otherwise).

As we have seen, Genesis explains that human beings are dust with a role and a relationship. But here's the problem: the original audience was a race of people enslaved by a powerful nation that believed in lots of powerful gods. How could they understand that there was actually only one God and that the one true God wanted a special relationship with them?

They were enslaved. They were experiencing all sorts of undeniable evil. Their God must either not care about them or not be powerful enough to help them . . . Or maybe there was more to the story.

Moses explains that in the beginning, after God created the heavens and the earth and all the stuff on the earth, he planted a garden and put Adam there to tend to it. He formed Eve to join Adam so that together they could bear God's image in the garden. Adam and Eve knew God and his goodness, but they didn't know evil. They didn't know what God was not.

God was very clear with Adam and Eve. He told them that they really didn't want the knowledge of good and evil, because if they tasted the fruit of that tree (quite literally) they would taste death. And we know how this story goes, right? They really wanted to know, so they broke the command, ate the fruit, and very soon afterward suffered the excruciating pain of being kicked out of the garden and then having one of their sons murder the other. They experienced what God was not. They committed and suffered evil. "How do we turn it off?," I can imagine them crying.

But the story doesn't end there. They go on sinning and suffering and their descendents go on sinning and suffering until God puts a special call on Abraham to form a brand new nation that will eventually grow to bless all of humanity by bringing divine image bearers back into a right relationship with God.

> Adam and Eve knew God and his goodness, but they didn't know evil. They didn't know what God was not.

This is the point of the book of Genesis— the first book of the Bible: "I know you are suffering," Moses is saying to the Israelites living in Egypt, "but your suffering isn't as senseless as it feels. We are in the middle of a much larger story, and there is peace and purity for us in the future if we follow God's plan."

## 3. Abraham, Isaac and Jacob . . . and then Jesus

God formed human beings by picking out special bits of creation for a special purpose. He empowered them for that purpose by relating to them more closely. Then He formed a special culture for Himself by picking a family for a special purpose. He empowered Abraham and his family for that purpose by relating to them more closely: a covenant community with a role to fill, empowered to fill it by a special relationship with God.

God called Abraham to move from the outskirts of Mesopotamia to the heart of the Fertile Crescent— the land of Canaan. They put down roots there and formed a sense of national identity.

But they didn't stay long.

Abraham's son Isaac had a son named Jacob. Jacob clung to God for a blessing even though he knew he would suffer for it. So God changed his name to "Israel", which means "wrestles with God". Out of this struggle comes a special knowledge of good and evil. The people of God are clearly part of a process, being defined and refined. Then a famine came to the land and they had to move to Egypt, where they were an ethnic minority that was respected at first but later enslaved.

Finally(ish) God called Moses, a descendant of Abraham, Isaac and Jacob to lead all of the other descendants of Abraham, Isaac and Jacob back to the land of Israel (or Canaan). He wrote the first few books of the Bible (which basically just explain what we have covered so far) and off they went.

They had generations of struggle and rebellion, wandering and war. They eventually managed a fairly powerful three-king dynasty (Saul, David and Solomon), but they (and especially their sucessors) kept turning away from God, and without that special intimacy with God, the Israelites really weren't anything special.

Eventually, they were conquered for good. They were allowed to keep up their culture and (most importantly for our purposes) the Old Testament, but as far as physical kingdoms are concerned, they were mere subjects.

> *Out of this struggle comes a special knowledge of good and evil.*

But then, out of the dark malaise of this all-too-familiar bland ordinariness and oppression, one of those subjects— one of those descendants of Abraham, Isaac and Jacob— was born supernaturally. He was born out of special intimacy with God– just like the first human, only more so. In fact, that intimacy was eternal. He wasn't merely a divine image bearer, He is also the second person of the Trinity. His name is Jesus.

Jesus had a tough job. You might think being God would have made things easy for him, but that really wasn't his plan.

He came to the descendants of Adam, Abraham and Moses and offered sacrificial leadership. He offered to take the effects of sin— the physical experience of suffering— upon himself so that anyone who accepts him as their leader will not have to experience alienation from God for themselves any longer. The taste of death will be taken from their mouths and they will inherit eternal life.

So, quick review:

God carved his first image bearer out of the stuff of creation and empowered him and his descendants with a special relationship, empowering them for a special role in creation. But to know the difference between good and evil, the first man broke that relationship.

> *The taste of death will be taken from their mouths and they will inherit eternal life.*

God carved Abraham out of the peoples of the ancient world and promised that all nations would be blessed through his descendants. That blessing has come in lots of forms, but most specifically in the form of Jesus, the eternal son of God.

## 4. The Apocalypse, Millennium and Reformation

Jesus came into the world at a specific point in history. His reception was pretty rough. His start was humble, to say the least. As a young adult, he drew some pretty decent crowds from time to time. He taught them well. But eventually he was rejected. He suffered and was killed.

But as we know, Jesus' death was not the end of the story. Not even close! After Jesus was killed, God made his power to save clear for all to see. By the power of the Holy Spirit, the Father raised the Son from death and the grave.

And after His resurrection, Jesus showed people that He was alive. And He was not just a disembodied spirit! He was (and is) a bit of enlivened dust. He, that precious, specific bit of dust, died and was placed in a tomb. But then, three days later, that very same bit of dust was resurrected. The tomb was emptied. That event, now called "Easter" is celebrated by Christians as the most important event in human history.

After His resurrection, Jesus was physically taken up into heaven with a promise to physically and actually return very soon to complete the hard part of bearing the divine image— filling the earth and subduing it with the love and the justice of God— by removing all traces of evil.

We ate the fruit of the tree of the knowledge of good and evil. We willingly tasted evil and death. "How do we turn it off?," we groaned. "Who is able to drink this cup to its dregs?" "I can, and I will," answered the only One who could– God Himself. "Now follow me," He said to us, "and when I return you will never taste that evil stuff again."

That is a quick summary of the cosmic history revealed and resolved by Jesus' life, work and resurrection. But in regular old human history, things have taken a little longer to play out.

After Jesus ascended to heaven and sent His promised Holy Spirit, things didn't immediately get better on earth. In fact, they got worse. God's plan was fully revealed in Jesus, but that revelation ("apocalypse" is the Greek word for it) was, for at least the first few centuries, nothing short of apocalyptic.

Jesus had been sentenced to death, and following him was not allowed either. His early followers were crucified, burned alive, grilled on iron chairs and fed to wild animals for entertainment. It was bad. Very bad. For decades. For generations. For centuries. But it wouldn't be this way forever. Turtullian, an African

> God's plan was fully revealed in Jesus, but that revelation was nothing short of apocalyptic.

Christian who lived in the second century and argued that the Roman Empire should stop persecuting Christians put it this way, "the blood of the Christians is the seed of a new life." This was the pain of the apocalypse. It wasn't until 312 AD that things began to change.

After the (nearly three centuries long) apocalypse came "the millennium", the 1000 year reign of Christendom . . . more or less.

Jesus came into the world at about the height of the Roman Empire. After his death, Christians were persecuted and executed in all sorts of horrible ways. But by 300, Roman imperial rule wasn't going so well either. It had been divided into Eastern and Western Empires, and Constantine was fighting for control of the West. Things also weren't going so well for him. In 312, he squared off against a much larger army in the Battle of Milvian Bridge. Although Christianity was illegal, hoards of lower class folks were committed Christ followers. Many of them were in Constintine's army. So the night before the battle, Constantine made an alliance with Christianity. He had his soldiers paint Christian symbols on their shields and they went off to win a decisive victory—ushering in a full millennium during which Christianity was the dominant religious philosophy of western culture.

When Christ returns in glory, he will unite heaven and earth and reign forever. It will be awesome . . . but this was not that. Not even close. From 312 to around 1312 (more or less), Christianity was legal and even respected as the most trusted source of truth. But the reign of Christ and the reign of Christendom are definitely not to be confused. It is probably safe to say that the millennium in question was more "churchy" than "Christian".

That's not to say the reign of Christendom was all bad. Many in the modern world are too dismissive of the Millenium— calling it "The Middle Ages" (between the "cool kids" of the Classical period and themselves), or even just "The Dark Ages".

But two of the greatest thinkers the world has ever produced lived in North Africa in the 300's. Among other things, Athanasius helped us understand the divinity of Jesus and Augustine helped us understand the relationship between time and eternity. Without them, we couldn't consider the eternality and trinity of God (mentioned above) settled concepts in the Christian world. And it was Athanasius who first listed out the 66 books that comprise the Bible.

> Centralized Christian authority has been in fairly steady decline for the better part of a thousand years.

Augustine was trained in the philosophical tradition of Plato before converting to Christianity, so throughout his writings we see him grappling with questions about the soul, when life begins and the importance of the resurrection. Then, toward the end of the millennium, we have Thomas Aquinas, who revived the study of Aristotlian philosophy. His work not only corrected a lot of over-reliance on Plato but also paved the way for the scientific revolution— the advances of which are the source of most of our modern hubris.

Christendom wasn't that bad . . . but it wasn't the end of the world either.

Starting in 312, the Christian church is where the Western world went for answers to all ultimate questions. That worked pretty well for a long while. But with humans, absolute power almost always corrupts. The official Protestant Reformation, which pretty much ended the church hegemony of the "known world" for good, didn't come until 1517. But I think it is pretty safe to say that the Reformation was more a sign that Christendom had ended than the cause of it.

Most historians agree that the failed Crusades (1096-1291), the Plague (1346-1353), the 100 Years War (1337-1453), and rampant church corruption had folks looking for a new authority figure for generations before 1517. What's more, the Reformation didn't claim that Christianity wasn't the answer. It said that the truth of Christianity was taught more clearly and authoritatively in the Bible than the human authority structures of the central church powers. It definitely got the crisis of confidence out in the open, but it took a few years to figure out what to do about it.

After the Reformation and the slow but undeniable destruction of the millennial reign of Christendom, we entered into an era that has come to be known as "Modernism". After Modernism . . . well . . . liberal democracy.

We will talk a lot more about Modernism and liberal democracy in later chapters. For now, let's just note that centralized Christian authority has been in fairly steady decline for the better part of a thousand years.

First crashed the full authority of the Roman hierarchy (or "Roman Catholic Church"), then national or "state sponsored" churches, then major, trans-national Protestant denominations, then independent evangelical leaders who gained too much power and/or prestige. It seems like everywhere we look these last few centuries we see sin, scandal and corruption throughout organized Christianity. Religious authority just isn't what it used to be! But when it's fully gone, what'll be left? Is nuclear war all we have to look forward to? There's a happy thought.

Maybe the actual, physical return of Christ is the next major change we have to look forward to. Maybe when the reign of the worldly Christians has fully run its course (and shown its bankruptcy apart from special intimacy with God), Christ himself will physically return. Maybe that will be very soon. But many have thought that before.

World history seems to be breaking down. And yet it keeps going . . . and going . . . and going.

The truth is, world history still hasn't ended. In fact, as much as it might have felt like it was going to, not even Western culture ended with the Reformation.

The power structures were in shambles, and everyone who tried to rebuild them failed. But then, a century after Martin Luther and the Reformation, out of the broken and exposed bankruptcy of the millennial intellectual hegemony, came Rene Descartes, the father of Modern Philosophy.

We will talk more about Descartes in the next section, but not because he plays a very large role in the "big picture" historical introduction that has been the goal of this chapter.

Students of redemptive history will have recognised what we have covered in this chapter as Creation, Fall and Redemption. The only major epoch of redemptive history yet to be covered is Glorification. We have mentioned the return of Christ as something to be looked forward to because it is a major piece of history that must be kept in mind. We will discuss it several times in future chapters, but since it hasn't occurred yet, it will not be discussed with any more detail now.

> *The truth is, world history still hasn't ended.*

Remember: I am arguing that human beings are best understood as dust with a role and a relationship. But . . . we're not just our bodies. We are physical beings, but there's more to us than that. We are also minds and souls. Human beings are multi-faceted. And being human can get a little confusing sometimes.

In the next few chapters, I will argue that understanding what it means to be human requires understanding from the first, second and third person perspectives. Roughly speaking, these perspectives correspond to the mental, social/spiritual, and physical aspects of being human.

Sometimes we refer to our mental lives as "our minds" and our physical lives as "our bodies". Within the history of Philosophy, this way of talking led many to believe in some form of mind/body dualism– opening up the possibility of disembodied minds and the related problems of mental causation and what has come to be known as "the mind/body problem".

But if we reject this dualism, we don't have such problems. Talk of human "minds", then, is just talk about ourselves against the backdrop of ideas. Likewise, talk of human "bodies" is just talk about ourselves against the backdrop of other physical objects and talk of "souls" is a way of thinking about ourselves against the backdrop of our deepest, most strongly held beliefs and values. But we will

> *Understanding what it means to be human requires understanding from the first, second and third person perspectives.*

have much more to say about that in the next few chapters.

For now, wrapping up this extended introduction, I'll provide a chart we can use to roughly categorize some of these related ideas.  But remember: grammatical rules always (or at least almost always!) have exceptions.  The exceptions don't disprove the rules.  Language doesn't work that way (usually).  Exceptions just put the rules in context.

| Perspectives | | |
|---|---|---|
| **1st** | **2nd** | **3rd** |
| Mind | Soul | Body |
| I am a string of thoughts in the world of ideas | I am a unique set of values and relationships | I am an object in the physical world of objects. |
| I am a process | I am a concept | I am an object |
| Descartes likes me (as do folks who want to "download" their memories into a piece of hardware that can keep up "their" consciousness after they die) | Taylor likes me | Scientists like me |
| Constructed largely by my past intentions and attention | Constructed largely by my society | Constructed largely by my parents |
| I | You | It |

# Part 2: Perspectives

# Chapter 4: The First Person Perspective

## 1. The Rise of the First Person Perspective

Have you ever been lied to?

Of course you have. It stinks, doesn't it?

It really, really stinks when it's about something important. Or if the liar was really close to you— if you really trusted them deeply.

If you know the pain of having been lied to about something very important by someone you deeply trusted . . . I'm sorry. It shouldn't be that way. I bet you've spent a lot of time and effort trying to protect yourself from experiencing that pain again.

As we explained at the end of the last chapter, Rene Descartes lived his whole life in the shambles of the broken down power structures of Christendom. Intellectual life in those days was a minefield of faith, corruption, power and disjointed fragments of ultimate truth.

Descartes was a Mathematician. He longed to protect himself from the pain of doubt. He knew what it is like to be lied to by someone you trust about something of utmost importance. So he wanted nothing less than to bring the clarity and certainty of Mathematical proofs to Philosophy and, ultimately, to Theology.

If you've heard of Descartes, you've probably heard his most famous line, "I think, therefore I am." Do you know why he wrote that? It is from his short, powerful book, "Meditations on First Philosophy" published in 1641. He wrote it to prove that God exists and that the human soul could survive the death of the human body.

The book is a sort of narrative, describing his train of thought as he grappled for something to believe in. He starts by expressing his frustration with all the lies and uncertainty. He decides to methodically clear his mind of all the junk he has heard all his life— the stuff that, now that he has lived long enough to be frustrated by how often he has been deceived, he knows he can no longer trust.

Is there anyone, anything, any simple idea that can be trusted?

Pushing hard on this question, he finally gets to the one idea he found to be immune to doubt— an idea that, the more you doubt it, the more you affirm it: doubt itself.

See how his experience with Mathematics plays in here? As soon as you begin working with a variable, you begin to define it. Whether a given variable exists as anything more than a concept may be debatable, but as soon as the variable is given, it exists— at least as a concept. The more that concept is considered, the more often it is conceptualized, the more it exists— at least as a concept. If he doubts whether or not he is doubting, he perpetuates doubt. So there you have it: something certain and lasting, immune to skeptical attacks, doubt itself.

Next, Descartes asks himself what is required for this to be the case. Since doubt is a form of thinking and thinking requires a thinker, he reasons that, even if everything else he has ever thought or known may be false, it must be true that, at least while he is doubting, he exists. When he doubts, he is a thing that is thinking. The more he doubts, the more he affirms his existence as a thinker. So, well, here comes his famous maxim: I think, therefore I am.

From that short but pretty darn certain piece of reasoning, he goes on to ask what he can learn from there. Unfortunately for Descartes, this portion of his argument– the part he set out to prove to begin with– has not been considered quite as convincing.

Since doubt requires thinking and thinking requires a thinker, what kind of thing is this thing that is required for doubt to be? His answer: an immortal soul.

Pretty much by definition in those days, souls were understood to be the kinds of things that think. Since he thinks, he must be a soul. And since everyone knows souls are immortal and created by God, walah: He is immortal . . . and God must exist as well. God exists because he created us. We are immortal souls because we think. And (for good measure) God must have put us in a real, physical world because it sure seems like he did and God wouldn't mislead us that much.

Maybe I have overly summarized a few steps in his reasoning, but that is pretty much the gist of it. Like most arguments from definition, his argument really only convinced those who agreed with him to start with. But that bit about his own existence being the only thing he can really know for sure has had some staying power. Within the history of philosophy, it represents a sort of Copernican revolution. Instead of seeing human beings as creations dependent on the more fundamental existence of a creator (or even simply as organisms brought forth from preexisting progenitors), Descartes encourages us to focus on the ideas first. The idea of God's existence is derived from the more foundational idea of my own existence. So we– or more precisely, our own first-person experiences– are the fundamental concepts that ground belief in all reality.

And there is a certain reasonableness to this, isn't there? We absolutely know that we exist, even if we don't know very much about what we are. And whatever it is that we are, we must like it, because whenever we feel threatened, we resist.

He set out to prove the existence of God beyond all doubt. Instead, he shifted the center of human certainty away from belief in God to the undeniably of individual human experience. Oops.

But it's OK. Descartes may have set Western philosophical history on an overly egocentric course, but the first person perspective really is a pretty big deal. It was probably undervalued in the past. Whether we know what our essence is all about or not, it is definitely true that we exist and we want to continue doing so.

But how do we know how to survive with such limited knowledge about who or what we are? Modernism is pretty angsty, isn't it? Maybe that is why, starting very soon after the Cartesian revolution into Modernism commenced, the effort has been underway to press on into some kind of Postmodernism. But let's not get ahead of ourselves.

> *The idea of God's existence is derived from the more foundational idea of my own existence. So we– or more precisely, our own first-person experiences– are the fundamental concepts that ground belief in all reality.*

## 2. Survival and the First Person Perspective

As we are now (perhaps somewhat painfully) aware, we have a first person perspective. And we kind of like it. At least we can't seem to help wanting our first person perspectives to continue. This is at least part of what our desire to survive (even in the face of death) looks like. We want our first person perspective to continue– maybe even beyond death. But . . . what does that even mean?

Have you ever almost died? I certainly have. Lots of stories come to mind. One of the most recent was a horrible crash on the Interstate that I narrowly missed.

As I drove through a major city, I was aware of a car a little ahead of me in the lane just to the right of mine. We were driving along at around 70 miles per hour, and I was thinking hard about something other than driving. The next thing I knew, something flashed around the car to my right. I heard a "bang", and the car disappeared. I was shocked! I looked in my rear view mirror to see a multi-car mess with a pick up truck airborne above it all. Then I went around the bend to the point where a concrete barrier obscured my view.

Wow! What just happened? As I drove along, wondering what I might be able to do to be helpful, I started to piece together what must have happened. I had just narrowly missed . . . something horrible. There was nothing I could do about it, but at least one of the people I had been driving alongside only moments ago was probably dead. It so easily could have been me. What would that have been like? With very limited information, my very active imagination started filling in gaps.

Can you imagine actually dying? From a first person perspective, what would it be like? Can you imagine? It may be harder to do than you think.

Let's try. I'll change my story a bit and see what happens.

Let's say I'm driving along at around 70 miles per hour, thinking about the relationship between inflation and mortgage interest rates or some equally puzzling mental pass-time, when my front bumper collides with some unknown, immovable object.

Time seems to slow down.

I see my hood slowly crumple as my car continues to plow into the object.

I am no longer thinking about the economy.

My foot reflexively begins moving from the gas pedal to the brake, but I don't even have time to notice. I see my airbag deploy from the center of my steering wheel, and everything goes white. Then, as my eyes are destroyed by the impact of the immovable object, I lose even the sensation of light.

> *Can you imagine actually dying? It may be harder to do than you think.*

What do I turn my attention to next? Hearing? Gone. Touch? Gone. Indeed, all sensation stops as my nervous system is destroyed by the violent impact.

Surely I would have lost consciousness by now. But what if I could, somehow, manage to think? To maintain a first-person perspective? The loss of sensation is easy enough to imagine. I can imagine a train of thought that goes from interest rates, through a quick crash, to complete sensory deprivation. I would probably describe it as "everything going numb." But what about the loss of other reasonably well known first person experiences of neural activity?

It is easy enough to imagine the loss of verbal ability. I have certainly been at a loss for words before. Imagining what it would be like to lose my ability to do math or recall memories also seems possible. But what if I lost all of these abilities at once? What would it be like to experience complete nervous system shutdown? Is it even possible to imagine what it would be like to be a corpse? It would probably feel like . . . nothing.

Maybe the corpse would cease being me upon complete nervous system shutdown. Aside from the ethical implications of my corpse losing its special place in creation, what would happen to my first-person perspective? If "I" am my first person perspective and my corpse were no longer me, what would be me? Could a bare soul take over somehow? Could I somehow regain consciousness without a functioning brain? Could I see again without functioning eyes? Could I hear, or feel or even have any sort of spatial awareness? If so, why can't we get our "bare soul" functions to take over for living people who suffer sensory or other neurological injuries?

These aren't very comforting questions. But surely losing consciousness is not quite the same thing as losing all first-person perspective. After all, I go to sleep more or less daily without fear of personal annihilation. Maybe in that case my dreams should be considered a sort of "intermediate state" of first-person perspective. But that doesn't really help me here, because sleeping doesn't cause complete nervous system shutdown– just an alternative and fairly well understood cycle of brain activity. Even my watch can track the cycles, and its interpretation correlates pretty well with my own memories of them.

Maybe, immediately upon death, my consciousness is transferred from my physical body (brain with all its various lobes, eyes, ears, etc.) to another body capable of sustaining something like my prior state of consciousness. Maybe this other body is somehow separable from my corpse. I am struggling to imagine what this might feel like, but . . .

I have heard various accounts of out-of-body first person experience, haven't you? But most of these certainly sound more like dreaming– more like involuntary acts of imagination– than true out-of-body experience. After all, if one didn't have eyes, how could we call it seeing? Could you see through other objects? In the dark? Around corners? Why or why not? And the same goes for other types of neurologic activity normally required for conscious activity. They just don't seem to work without embodiment of some kind or another.

> *It is probably easier to imagine what it's like to be a corpse (or a poached egg, for that matter) than to imagine what it would be like to be God.*

We might remember that God sees us, hears our prayers, etc. without having a body. But that hardly helps us because we know that his "seeing" and his "thoughts" are so drastically different from ours. He isn't limited by perspective, subject to surprises or .

. . well . . . spend much time thinking about it and you'll invariably come to the conclusion that we can't share his first-person perspective very well, even imaginatively.  The distance is just too great.  Although it is still quite a stretch, it is probably easier to imagine what it's like to be a corpse (or a poached egg, for that matter) than to imagine what it would be like to be God.  Unless, of course, we are trying to imagine what it would be like to be Jesus, who humbled himself to the point of sharing an experience like ours so that he could redeem and lead us.  But then, Jesus clearly had a body.  That's how we can imagine his perspective.  So that doesn't help.

## 3. The Bible and Disembodied First Person Perspectives

But are there any cases we know of in which human consciousness *has* taken place without a human body?  Well . . . maybe in the Bible.  Or at least maybe/kind of.  Let's look a little closer.

The first case I know of is recorded in 1 Samuel 28.  Samuel dies, then Saul wants to know whether he will be blessed in battle.  God used to answer that kind of question through Samuel, but now . . .

Desperate and afraid, Saul enlists the help of a medium from Endor to bring Samuel back from the dead so that he can speak with him.  Weird story, right?  But if you believe in the Bible as strongly as I do, you know it actually happened.  So let's look into it and see what we can learn.

Unfortunately, we don't get a lot of detail regarding the physical mechanics of the whole medium process or how it works exactly, but a few things are very clear from the passage. First, it is bad. Saul knew it was wrong, and Sammuel definitely isn't happy about it either.

More to the point, Sammuel seems to have been provided with some sort of temporary body for the occasion. The medium describes the body as "wispy" or "ghostly" and then "an old man, wearing a robe". It does seem that Samuel has a first person perspective, but it doesn't seem that this experience is disembodied– just . . . poorly embodied. His experience is mediated by the wispy, old man-ish thing wearing a robe that the medium saw.

> *After all, if one didn't have eyes, how could we call it seeing?*

Creepy.

Thankfully uncommon.

Probably a human first-person perspective without what we would normally call a human body, but not really a disembodied first-person perspective. And definitely not an experience that we should hope for in the face of bodily death! Maybe we should look for other examples.

Something similar could be said regarding the Mount of Transfiguration, described in Matthew 17. Once again, we have people who have clearly died

(Moses and Elijah) seeming to resume a first person perspective without any mention of their human bodies having been resurrected. But, once again, their experience is clearly mediated by some sort of physical body.

Although they are definitely taken aback by the encounter, it is clear to the disciples that Moses and Elijah are not disembodied souls. They even go so far as suggesting that they build a shelter and invite them to stay.

Once again, the whole story is shrouded in mystery. Jesus tells the disciples not to say anything about it until after his resurrection. Lots of aspects of this story are unknown. The one thing that is clear is that the people having whatever first-person experiences were had were not immaterial (if they were, why would the disciples want to build a material shelter for them?), so no help there either.

While not quite stories of first person experiences after death, two more biblical examples are pertinent to this discussion. The first is Paul's repeated use of language that seems to imply that he expected disembodied consciousness after death. The second is the thief who was crucified beside Jesus. Paul's contemplations are much more difficult to understand, so let's discuss his case first.

In 2 Corinthians 5:8, we read, "Yes, we are of good courage, and we would rather be away from the body and at home with the Lord."

As in our discussion of the creation of Adam in our historical introduction, we must acknowledge that if this passage is read with traditional, Platonic presuppositions (thinking of people as essentially immaterial souls who temporarily inhabit bodies during their lives on earth) this passage would appear to support those presuppositions. It sounds like what Paul is saying is that he would rather be a disembodied soul in heaven than go on living a life entangled with the physical matter of a human body.

> *The people having whatever first-person experiences were had were not immaterial.*

This reading seems to be further supported when cross-referenced with Philippians 1:21ff. In that passage, we read:

> For to me to live is Christ, and to die is gain. If I am to live in the flesh, that means fruitful labor for me. Yet which I shall choose I cannot tell. I am hard pressed between the two. My desire is to depart and be with Christ, for that is far better. But to remain in the flesh is more necessary on your account.

Maybe Paul really was a dualist. Most scholars have certainly read these passages as "proof texts" for some sort of intermediate state between death and resurrection in which first person experience can be had without brains, eyes and . . . well . . . really without any sort of body in this dimension.

I don't want to argue that this reading of these texts is completely unwarranted. Very reasonable, educated people– most of them throughout history, in fact– have come to that conclusion. But there is much more going on in these passages, and I don't believe the dualist reading is the best one.

Although it would take us too far from our current discussion to investigate here, certain other passages have also been taken from Paul's letters that, taken out of context, lead many to believe that Paul was a misogynist. I don't think he was. There are a few passages of his writing that make me scratch my head a bit, but I think explaining his belief in the equal value of male and female divine image bearers is both possible and important (and has been done by others elsewhere). I am a little glad such a discussion would take us too far off task for this work, however, because it is much more difficult to explain Paul's statements that seem to support misogyny than those that seem to support dualism.

But if he's not a dualist, what does he mean when he talks about being "absent from the body" and "present" elsewhere? To answer that question, let's look at one more brief passage. In Colossians 2:5, he's using that language again. He says: "For although I am absent in body, yet I am with you in spirit". If we read this through the lens of substance dualism, it sounds like his body is one place (absent from Colosse) and his spirit is in another (there with the Colossians). Is this the example of disembodied first person experience we have been looking for? Let's think it through.

Is he saying that his body has fallen into some sort of trance, allowing him to leave his body and travel as a bare soul to another city? That's what the old substance dualist reading would suggest. What about his first person perspective? Is it with his spirit, hovering around Colossae, or with his body, back wherever, moving his pen across the page? Could it be in both places at once? If his body is in one city writing the letter and his spirit is in Colossae with the folks he is writing to, where is the "real" Paul, with the body or the spirit? If we say Paul's spirit is the "real" Paul, which is what most dualists believe, then Paul didn't really write the letter. His body did it while Paul was out of town! The dualist reading clearly doesn't work here. It seems much more likely that what we are reading here is more a figure of speech.

When we read a little more of the context of 2 Corinthians 5:8, it becomes clear that he's using a similar figure of speech there— not exactly the same, but close.

> *I don't believe the dualist reading is the best one.*

If you haven't already consulted a commentary or study bible for an overview of Paul's argument in 2 Corinthians, this would be a great time to do so. In particular, look for a "big picture" outline of the letter and the position of 2 Corinthians 5:8 within that outline.

Paul's extended argument in 2 Corinthians is very familiar to New Testament readers. He is doing ministry– applying the "new reality" of the church to specific contexts his readers are struggling with– and putting his ministry in the context of redemptive history. In particular, he is connecting his ministry as a preacher of the good news of Jesus to God's work in the Old Testament through the law and Moses.

In chapter 3, particularly in verse 7 and following, he reminds his readers of the leadership provided by Moses in the desert (recorded in Exodus 32-34). In the desert, Moses met with God 1 on 1, then came back to the people with clarity about what God wanted from them. The people received that message in the doorways of their tents, and they were commanded (in elaborate detail) to set up a special "tent of meeting" for Moses and God.

That tent of meeting, later known as the tabernacle, was really just a tent. A temporary, mobile shelter. But they were told very clearly that they wouldn't be wandering in the wilderness forever. They would eventually have a lasting home and a temple with foundations fixed in Jerusalem. The tabernacle, or tent of meeting, was eventually replaced by Solomon's temple.

Solomon's temple wasn't just a tent. It was a temple. It was a building with foundations, built for people planning to stay around for a while. But even that home wouldn't last forever.

In John 2, we read of Jesus' zeal for the purity of God's dwelling place on earth. When people saw what he was doing (overturning vendor's stands and raging around with a whip) and asked who he thought he was to be taking such matters into his own hands, he replied, "Destroy this temple and I will raise it again in three days." Clearly, Jesus was teaching that his resurrected body was an even more permanent replacement for Solomon's "temple with foundations".

The tabernacle was a temporary structure that foreshadowed the temple. The temple was a worldly structure that foreshadowed Jesus' resurrected body at the center of the new heavens and the new earth. This image is right at the center of the story of the gospel– the story of God coming to dwell with his people without compromising on

> Solomon's temple wasn't just a tent. It was a building with foundations, built for people planning to stay around for a while.

the line between good and evil; the sacred and holy on the one hand, fully redeemed and sealed for all eternity and the profane and sinful on the other, cut off and sealed for eternal destruction.

In 2 Corinthians 5, Paul is building on this image. He pleads with his readers not to allow their sense of self– the ground of their personal identities– to be tied to this tent-world– the world that is passing away. They are to ground their identity in the resurrected Jesus.

When they tore down the temple of Jesus' body, it was raised again in three days. His resurrected body (and those of us who are united with Him) are a kind of firstfruits of all He created. We just have to put off the "tent world" and put on the "temple of the New Jerusalem world". Paul is telling the Colossians (and us!) to put off the practices of worldliness. The practices of sin and death. The practices that tie people to the current state of affairs. Instead, Paul is telling his readers to look for the New Jerusalem that is even now being prepared for us in the heavenly realms.

Paul knows that true life is to be found in Christ alone. He is in Christ, but remains active in the flesh– in the events of this world. He is united with Christ, and is thus already active in the world to come. But until Jesus' second coming, that future world will remain just that– future. As he wrote 2 Corinthians, he was painfully aware of his continued involvement in ancient world history. Somewhat ironically, he was a professional tent maker (literally making temporary housing) in the tent world, even as his union with Christ anchored his work as an apostle in the immortal world to come. When he dies, he won't "exit" his tent as a bare, disembodied soul as the dualistic reading would have us believe. He makes this point explicitly– twice in as many verses, saying: "if indeed by putting it on we may not be found naked" and "not that we would be unclothed, but that we would be further clothed."

> *Put off the practices of worldliness.*

Now let's return to the Phillipians passage we mentioned earlier and ask whether this reading offers us a better way of interpreting Paul's words than the traditional, dualistic reading. When Paul says, "My desire is to depart and be with Christ", the dualistic reading sees Paul talking about departing to be with Christ and asks, "where is he talking about going?". But I believe this is the wrong question. The right question isn't "where?" but "when?"

There are lots of passages we could turn to to illustrate this point: that Paul thinks of departing and being with Christ as a temporal thing more than a spatial thing. Maybe one of the clearest is back in Collossians– particularly the first part of chapter 3. He uses both "where" and "when" language in this passage, but it's clear that only the "when" language should be taken literally:

> Since, then, you have been raised with Christ, set your hearts on things above, where Christ is, seated at the right hand of God. Set your minds on things above, not on earthly things. For you died, and your life is now hidden with Christ in God. When Christ, who is your life appears, then you also will appear with him in glory.

For those of us who are in Christ, to be absent from the body is to be present with the Lord. But that isn't a statement of where we will be. It's when we will be.

That may sound a little overly rhetorical, but it shouldn't. We are definitely not talking about

75

anything mystical and mysterious here. Sometimes it is a little difficult to get out of the old, dualistic ruts of thinking, but it doesn't need to be confusing.

> *The right question isn't "where?" but "when?"*

Maybe things will get clearer if we bring this "where/when" distinction back to our "first person perspective of death" thought experiment.

There I am, driving down the interstate. I hit the unseen, immovable object and time seems to slow. The hood crumples. My foot jerks toward the brake. I barely have time to panic before all sensation, all brain function, everything I would call "thought" or a "first person perspective" stops.

But only for an instant.

I'm a bit disoriented, but also somehow feel more awake and clear-headed than I have ever felt.

Is that a car horn? A siren? Some sort of trumpet?

I hear the echoes of a blaring horn as I soar upwards through cloth, then wood, then concrete, then dirt. I plow through at least 6 feet of solid earth as if it were water.

I'm amazed as I realize the ground is every bit as solid as I knew it to be before— It is just that my body is now more solid, more permanent, more real than the ground. I must have died when my car hit that crazy immovable object. But before I even had time to panic, my attention was captivated by this new reality. I am not dead. I am more alive than I have ever been before.

But I am in a graveyard, coming out of a hole in the ground. The ground is still shaking a bit, and there are hundreds of beautiful, shimmering, horrifyingly powerful but also peaceful and holy people all around me.

With morbid curiosity, I wonder if my body had been embalmed. My fingers jump to the place I knew the mortician's needles would have penetrated my flesh.

Scars.

The marks are there, but the wounds have healed more beautifully than I could have imagined. I look to the others around me. They have scars too. Like badges of honor, I can see various signs of past suffering now shining in indescribable glory.

As I turn toward the being closest to me, who just emerged from a hole next to mine, I recognize a scar. I gave it to him. Or at least the origin of it. With a sharp twinge, I remember the harsh words I said to my oldest son about his grades just before I left for the work trip that ended my life.

But the sharp twinge isn't guilt. I know I have been fully forgiven. Actually, it is much more than just a twinge. Just five thoughts ago— back in two thousand . . . whatever— I would have called it completely overwhelming. But it isn't guilt. It is awe.

How could he have heard those horrible words— the last words I would ever say to him on earth— and had something so beautiful come from that dreadful experience?

As that sense of awe and wonder and joy overwhelms me, I notice the shimmering light that bathes us all. My fondest memories of bright, warm sunshine seem damp and cold in contrast to this new reality. I look to its source and see, not a burning ball of fire in the sky, but a person. It is a human body. But no mere human. He is the light of the world. We didn't really understand that before, but . . .

OK. As much as I would love to continue this thought experiment, I think you get the point. There is no time, from the first person perspective, between death and resurrection. It is not that we are waiting, or sleeping or . . . definitely not sitting around on some cloud playing a harp. Time just doesn't work that way.

Remember, back in the last chapter, when we said that time isn't a substance? That's why we can say that there is no time between death and resurrection. There is no person between death and resurrection. There is a corpse, and we will talk about that more in another chapter. And there is the memory of the person. We will talk more about that later too.

For now, let's just make this point very clear: people who are dead right now are not alive right now. They aren't alive in heaven right now. They aren't alive, sleeping in their graves right now.

They aren't living as ghosts in their ancestral home or something else creepy like that.

If a person is dead right now, they are not alive right now. They were in the past. They will be in the future (when Christ returns). But as for now, their spirits—their lives—have returned to God the Father who gave them (Ecc. 12:7). The most interesting sense in which they exist right now is in the mind of God– not as conscious or unconscious/comatose people waiting, but as memories, held by God in such detail that when the dust of their corpses is transformed into the new, glorified, uncursed dust of the new earth, God can resurrect them in full and perfect detail.

While we can't know this for certain, the Biblical account of death and resurrection would indicate that if a person was trying to solve a math problem when they died, God will remember and resurrect every synapse in their brains in such detail that they will be focused on that math problem at the resurrection (though, as we saw in our little thought experiment earlier, they may quickly be distracted by subjects even more fun than math!).

*There is no time between death and resurrection.*

This understanding of time, death and resurrection also helps us make sense of one additional Bible passage that is often misunderstood: Jesus' statement to the thief on the cross next to him as he died. In Luke 23, verses 39-43 we read:

One of the criminals who hung there hurled insults at him: "Aren't you the Messiah? Save yourself and us!" But the other criminal rebuked him. "Don't you fear God," he said, "since you are under the same sentence?" We are punished justly, for we are getting what our deeds deserve. But this man has done nothing wrong." Then he said, "Jesus, remember me when you come into your kingdom." Jesus answered him, "Truly I tell you, today you will be with me in paradise."

The traditional, dualistic reading of these verses has resulted in a quagmire of "Biblical" metaphysics. I won't waste time with various attempts at answers, but the internet is full of some doozies!

What is the relationship between Christ's kingdom, heaven, paradise and the new heavens and the new earth? When do souls go where? What about bodies? How is Jesus in the grave and paradise later that day? And does he go preach in hell along the way somehow??

Understanding that time is just a way of measuring relationships– a concept, not a substance– clears away all this confusing nonsense. "Today" simply means "within the next few hours you experience". The first person experience of the thief has no "gap" between his death in approximately 33 AD and his resurrection significantly over a thousand years later.

Remember the twins paradox from high school physics? Because the physical world has a universal maximum speed, if one identical twin jumps on a rocket ship and flies quickly around Pluto, the twin who stayed on earth will be much older than the twin who traveled. It is a fun paradox, but the lesson is clear– time isn't a universal substance. It isn't itself a "thing", it simply tracks relationships between things. So, as Augustine mused over 1,000 years ago, we must distinguish between the different things we might mean by "time". After all, time-for-God– who understands the past, present and future simultaneously and with absolute clarity– is very different from time-for-us. As our first person perspective thought experiments have shown, we must also distinguish between time-for-us and time-for-me– especially if we are to make sense of Biblical descriptions of death and resurrection.

> "Today" simply means "within the next few hours you experience".

So when Jesus was talking to the thief, he said "today", meaning "in the next few hours you experience." If he had been talking to the Roman soldiers, Simon from Cyrene and the women who had followed him from Galilee, he might have referenced the same future point in time according to the "time-for-us" perspective, saying "centuries and centuries from now, when I return in glory and the dead are resurrected, this thief will be with me in paradise– which is my kingdom, which is the new earth which will be united with the new heavens."

But to a new convert being tortured to death, "today" was probably a much more true and comforting way of saying it. For the thief on the cross in the throes of death, and (as we saw earlier in this chapter) for philosophers of the modern era, the first person perspective is what matters most about personal identity in the face of death.

But we all know deep down that personal identity is about much more than just that.

As we will discuss in the next two chapters, the second and third person perspectives matter too.

# Chapter 5:  The Second Person Perspective

## 1. A Quick Philosophical History from the 1600's to the early 2000's

In the last chapter, we saw how Descartes managed to fabricate a surprisingly sturdy bunker of mental assurance from within the fractured shambles of religious authority. Even if nothing else was firm and lasting in the long list of things we are encouraged to believe, he argued, because we cannot deny that we are thinking (ie. that we have a first-person perspective) we can know that we exist. Upon this seemingly indubitable foundation he, and many after him, attempted to build systems of philosophy.

For the next 300 years or so, the philosophical system known as "Modernism" reigned supreme. There were many different schools of philosophy and many great thinkers; many denominations of Christianity and many great preachers, but all Modernists pretty much agreed on two things:

1.  A human being is essentially a first person perspective (or mind) and

2.  God (more or less the God of the Bible) plays a pretty big role in how the world is to be properly understood.

By the late 1800's, however, Modernism was showing serious signs of wear.  On the one hand,

Modernism's extreme focus on the individual human mind produced enormous progress in Math, Science, and eventually Engineering. On the other hand, however, God seemed less and less involved in a profitable understanding of the universe– so much so that in the waning years of the 19th Century, Friedrich Nietzche famously declared the death of God in Western culture.

In many ways, the world wars of the early 20th century can be understood as the brutal, ugly and unbalanced crescendo of Modernism; the inevitable consequence of the unchecked ideological supremacy of the ego. The thinking thing, the human mind, the one, sure, undeniable existence demonstrated its power and control of the world through increasingly sophisticated Mathematics.

> Could the very concept of personhood survive the death of God?

Range calculations, supply chains, and eventually the esoteric logical extremes of quantum physics all seemed to matter like they had never mattered before.

Cathedrals crumbled.

Cultures capitulated.

The patience of God, like everything else known to humanity, was tested to the breaking point.

In the end, the awesome power that reigned down from the heavens to end the war was shockingly different from what we had come to expect from the Judeo-Christian God. If God was still at work in the world, it seemed that He was no longer revealing His will through word and sacrament. The ultimate power in the universe didn't look like that any more. The face of God, it would seem, was now most completely revealed in the seemingly limitless marvels of human technology.

Pretty dark, right? Well, I think we can safely say that, as far as shedding light on our sense of human identity goes, the world wars were an unmitigated disaster. But that's OK. We can learn about as much from failure as success, right?

After the wars, debates between Capitalism and Communism ravaged most of the world's intellectual energy. These debates were essentially debates about how multiple first-person perspectives could work together most effectively.

Simultaneously, a more startling, less political question began circulating among philosophers and artists. Could the very concept of personhood survive the death of God? If we could no longer believe in something like the Judeo-Christian God, could we still cling to something like Cartesian Individualism? Could the first person perspective or any other rational sense of the human self survive?

This question rose to prominence in Europe with characteristic daring, cynicism and recklessness.

This line of questioning and the grasp for answers came to be known as "Existentialism".

Existentialism started at the end of the 19th century with the Danish thinker Soren Kierkegaard. While he still believed in God, he spent a lot of time exploring an idea that he (at least sometimes) seemed to actually believe: radical individualism– the idea of the absolute primacy of the first person perspective. With stark, bold and vulnerable frankness, he highlighted the courage required to embrace individualism's loneliness and anxiety.

As seems fitting, most of his writings were in the form of private journals. They are lengthy, honest, and complicated; deeply intriguing and notoriously difficult to understand. While Kierkegaard is now known as the father of Existentialism, this system of thought reached its apex in the works of Sartre and Camus, two brilliant, creative, thoroughgoing atheists living and writing against the tumultuous landscape of post-war France.

In 20th century America, the existence of God wasn't directly denied– at least not as often or as fashionably as it was in Europe. Instead, discussions of God were sequestered behind the iron wall dividing church and state, or "personal" and "professional" life.

Building on the late 19th century work of C.S. Peirce, William James and John Dewy developed a system of thought they dubbed "Pragmatism". If it is true, it works. And if it works, it is true. Likewise, God may be working privately and personally, but as for objective and public reality, he is rarely a force to be reckoned with.

As America ended the 20th century as the almost undisputed world superpower, so did Pragmatism. This system of thought was most fully developed toward the end of the century by Richard Rorty, but it never quite captured the public imagination as much as the European school of Existentialism– probably because its tenants are less noticeable. The ideas did not shock us or challenge us. In fact, they have become more or less ubiquitous.

As we approach the end of the first quarter of the 21st century, it seems that this "Post-Modern" line of thinking– asking whether Cartesian-like certainty in the existence of the self can exist without a correlated belief in something like the Judeo-Christian God– is beginning to settle into a more optimistic (if more totalitarian) resolution.

This movement, which started as a dark, alarming, and stark admission may have been successfully calmed and made more popularly palatable by the characteristically nice Canadian philosopher, Charles Taylor.

In his seminal works, "Sources of the Self", "The Malaise of Modernism", and "This Secular Age", he argues affirmatively: Yes, a sense of self can survive the death of God. But for the human sense of self to survive, we are going to need to work together.

Whereas Descartes said "I think, therefore I am", Taylor says each of us are who we are known to be. But whereas the Existentialists put the burden of personal existence on the individual, Taylor shifts this burden to society as a whole. The primary role of a liberal democracy, he would argue, is to create and sustain a moral fabric upon which individuals may define and explore their identities relative to others.

> *The turn from Descartes to Taylor as Modernism has developed is best understood as the turn from the primacy of the first person perspective to the primacy of the second.*

Although his works are exceedingly thorough and notoriously long, Taylor explains the situation succinctly in this paragraph from a lecture he gave in 1994:

> We become full human agents, capable of understanding ourselves, and hence of defining our identity, through our acquisition of rich human languages of expression. For my purposes here, I want to take language in a broad sense, covering not only the words we speak, but also other modes of expression whereby we define ourselves, including the "languages" of art, of gesture, of love, and the like. But we learn these modes of expression through exchanges with others. People do not acquire the languages needed for self-definition on their own. Rather, we are introduced to them through interaction with others who matter to us– what George Herbert Mead called "significant others". The genesis of the

human mind is in this sense not monological, not something each person accomplishes on his or her own, but dialogical.[1]

While there is much, much more that could be said about these past 400 years of Philosophy, we don't need to get too bogged down into detail. For our present purposes, I believe the turn from Descartes to Taylor as Modernism has developed is best understood as the turn from the primacy of the first person perspective to the primacy of the second.

## 2. Knighthood, Empathy and the Second Person Perspective

Do you know how someone becomes a knight? I am not sure about all the particulars, but I think the gist of it is that knighthood is conferred upon you by someone who has already been knighted. This, of course, raises the question of how the first knight came into being, but . . . I don't really want to talk about knighthood here, I am just trying to make a point about personal identity. So hold that thought. We will come back to it.

Have you ever thought much about the ethical treatment of animals? Why is it OK to eat some living things and not others? Don't worry, I don't want to dip us too deeply into that thorny, always emotional and often irrational discussion either, but . . . well . . . hold that thought too. Just for a

---

[1] See p.32 of <u>Multiculturalism</u>. Gutmann, Amy (ed.) *Princeton University Press;* 1994.

minute. I will tell a quick story that should tie things together.

From 2014-2016, my family attempted a little experiment. We tried to live as independently as possible. We bought a couple acres in the North Carolina mountains, I took a job less than 2 miles down the road at the small school we selected for our three boys and we built a mini-homestead.

My wife focused mainly on the house, the bees (she is a third-generation beekeeper), and the overall design of things. I focused on building, fixing and restoring things, as well as keeping enough wood cut, split, and stacked to heat our house through the chilly mountain winters. All of us worked together on the garden and (our favorite and most successful venture) the 29 chickens.

Those chickens were amazing! In addition to providing almost 2 dozen eggs per day, they each had different preferences, talents and tendencies. The boys not only fed and protected the chickens, they also got to know each of their personalities and quirks.

When the boys needed money, we allowed them to sell eggs to our neighbors and to the tourists that frequented the scenic drive that ran by our house. Always on the lookout for an extra dollar, our middle son, who was in 4th grade at the time, had an idea for an egg-sale special that proved hugely popular with the tourists and their grandchildren. He said that if a customer bought a dozen eggs at once, they could pick their favorite egg and he would catch the chicken that laid it, tell them all about the chicken and let the customer pet it.

Pretty cute, right? But what does this have to do with human identity? We are getting there.

Fairly early in our chicken-and-egg venture, we ran into rooster problems. Actually, it was a series of problems.

Our first rooster, Carolina, picked a fight with the dog. As you might imagine, that problem pretty much took care of itself. Carolina's departure from the land of the living meant a promotion for his second-in-command, Urchin.

Urchin was as tough as Carolina, but also smart enough to give the dog her space. His life and reign might have been fairly pleasant and uneventful if not for the youngest of our three roosters, Roo.

Through sheer physical strength, Roo couldn't really stand up to either Urchin or the dog . . . but he wasn't docile either. Mischief and mayhem were his constant, carefully chosen and constantly crafted companions. Somehow managing to find the last nerve of every living being within a substantial radius of his being, Roo was an absolute menace.

Fairly early in the process of Roo making his character known, I thought of an easy solution. This would be a great opportunity to teach the boys an object lesson in where food comes from.

We allowed little Roo to grow to at least the size of the chickens we saw in the grocery store. By then, he had annoyed all of us to the point that we were ready to see him go. I read articles on killing, processing and cooking chicken and we all made plans for how we would eat him.

As the days turned into weeks, we kept planning, preparing, and fantasizing about Roo's end. Urchin and the hens seemed to know what we had planned, and longed for the day as well. Roo kept relentlessly sneaking around, attacking out of nowhere before running off to safety, messing up the food and water when no one was around and generally getting on everyone's nerves. Couldn't we just off him already!? But we never found quite the right time to actually . . . do it.

Eventually, I realized what was happening. We weren't just longing for Roo's end, we were also avoiding it.

We couldn't do it.

Maybe we could have killed him, but I kind of doubt it. And when it came to processing, cooking and actually eating him . . . no way. We just couldn't do it.

Don't get me wrong, we could eat chicken. We did it all the time. But we couldn't eat Roo. It just didn't feel right. Annoying as he was, he was . . . a member of our community. We had named him. We knew his character. He wasn't just an animal. He had been . . . personified. And it is not OK to eat people.

So that is how knighthood, the question of what things it's OK to eat and personal identity come together. You can't eat something you refer to in the second person. I can eat *it* (a "third person" thing) all day long, but I can't eat *you* (a "second person" thing). And what is the difference between something appropriate to refer to in the second person and something we can safely sequester to the third? In a word, empathy.

But empathy doesn't just pop up out of nowhere. First of all, it requires a person. Then, it requires something else for that person to relate to. And the person has to choose to relate to the other thing as another person– not just an 'it', but a 'you'. See how it's kind of like knighthood?

> *If a member of a community confers second person status to something, that act has ethical implications for the entire community.*

Let's look at another example. While it may be a bit sappy for my taste, the movie "Castaway" from 2000 does a pretty good job of exploring this idea.

In the movie, Tom Hanks plays an FedEx employee (Chuck Noland) who survives a plane crash over the Pacific Ocean. The sole survivor, he washes up on a deserted island and manages to survive four years before making it back to civilization.

The second-most important character in the movie is a volleyball. Chuck finds the ball in a package that washed up on the shore. In a scene dripping with unexplained symbolism, Chuck Noland (You get it? "Chuck", like "to throw" and "Noland", like the muddling up of the words "noone" and "no land") tries to build a fire. Very "human person" thing to do, right?

The problem is that, in the effort to build a fire, he cuts his hand. Just his hand. Not both his hands and feet and side—like he had been crucified—that symbolism would be just too obvious. He only cut his hand . . . but he was mad about it. In his frustration, he directs his wrath at the volleyball.

What did the volleyball do, ask for the knowledge of good and evil?

No, but it had been cast (or "chucked") down from the heavens with Noland and that made it the most deserving thing around of such a personal thing as wrath.

When Noland, in his wrath, chucks the ball away from his presence, he accidentally smears the ball with his own blood.

I don't guess there is any symbolism in that. I mean, it is not like he breathes his own breath into it or anything. It's just wrath and blood and . . .

Anyway, seeing his own blood on the ball somehow quells Noland's wrath and makes him decide to knight the ball. Using his blood as a big, bright marker, Noland draws a face on the ball and it becomes his best friend. He no longer regards the ball with a third person perspective. He grants second person status to it . . . or him.  More than merely slating his wrath, this act abates his loneliness as well.

While Noland's actions seem somewhat understandable in the movie, he's also clearly going crazy, right?

See how we teach others to be judicious in their use of their ability to confer personhood?  We threaten to suspend, "laugh away" or otherwise deny their powers of personification.

Pick an animal or two to keep as pets and you are well within your rights, but if you go around extending a second person perspective to everything, we will begin to discount your sense of discernment.

After all, if a member of a community confers second person status to something, that act has ethical implications for the entire community.  It's not just wrong for me to eat my pets.  I can't eat yours either.  But if someone gets too liberal in their use of their powers of personification, we have to protect the community by putting that person's powers into question.

So let's go back to the knighthood example.  If a knight knights you, you are a knight.  But where did the first knight come from?

Maybe he acted heroically and sacrificially to save an entire civilization from certain doom.  Once the people were saved, they needed to honor him for his selfless act.  They gave him a title of honor, but that wasn't enough. They also all agreed that anyone he decided to share the title with (his sons, comrades in arms, maybe even his closest friends) would also be respected with the same title. As long as he didn't confer the title on too many people, that would probably work. Maybe that is how knighthood actually came about. Maybe not. Wherever the idea came from, it definitely works a lot like personification. The idea can have power and meaning only as long as neither he nor anyone else he conferred the title on misuses it.

But what if knighthood had a less noble origin. What if it was just some guy (we'll call him "Wayne") who made it all up. "Hey You," he said to a random stranger. "Kneel down and let me pretend I'm gonna cut your head off with this big sword and I'll give you special powers." Think of it that way, and knighthood starts to lose its luster.

If that's the way it all started, knighthood becomes the mark of a fool, not a hero. In fact, the mere possibility of knighthood being Wayne's idea seems to degrade the whole concept.

## 3. The Power of the Second Person Perspective in Historical Context

Personification is a beautiful, dignifying gift. Objectification, or the attempt to de-personify something, is . . . well . . . pretty much the opposite. But that doesn't mean that the decision as to whether or not to attribute a second person perspective is an all or nothing issue when it comes to dignity. Nor is it something we always all agree on— just look at the history of debates over abortion and euthanasia, for example.

But let's not get bogged down in controversy. Instead, let's see if we can gain additional clarity by reviewing some key lessons of history again.

First, let's remember that the Bible says the last stage of creation before the day of rest was when God personified a specific lump of dust; a bit of His creation for which He had very special plans. Then they rested together.

But it is important to remember that the special creation of Adam and Eve wasn't the first act of personification. It was the first one in all creation, but that doesn't mean it was the first. The eternal, triune God of the Bible is three persons affirming each other's personal identity for ever and ever and always.

But in the beginning of creation, God created the heavens and the earth. And then on the sixth day, He separated human beings from the rest of creation and chose to have an intimate relationship with them. He made humans His representation before the rest of creation by personally personifying us. And He's no "Wayne"! His act of personification

> *The eternal, triune God of the Bible is three persons affirming each other's personal identity for ever and ever and always.*

gave us the dignity we needed to do His will. We are dust with a role and a relationship.

And of course, we know how this big story goes. That act of divine empathy was begun at creation, frustrated by the Fall, perfected on the cross, solidified by the resurrection and ascension of Jesus, and will (eventually) be completed by His second coming. That's the "big picture" story of human personification . . . and pretty much the history of life, the universe and all creation in a nutshell.

So with that in mind, let's apply these lessons back to personhood and the last millennium or so of the history of Philosophy.

Back in the days of Christendom (roughly 312-1312 AD), the centralized church was the sole authority on personhood. If the Church leaders "knighted you" with baptism, you were a person. If you were an unbaptised pagan or excommunicant, you were less than fully human. We might not want to eat you necessarily, but your life wouldn't necessarily be protected, and you certainly wouldn't be allowed to take communion or be buried with the fully personified people.

As church authorities began disagreeing more and more, however, their ability to confer personhood status began to sound more like Wayne's version of knighthood. People began looking for a deeper sense of dignity, worth and assurance. For a while, church authorities were propped up by the physical power of the state, but in the end, that just made the whole problem bloodier.

Then along came our favorite Christian mathematician. By his day, the forms of dignity conferred by the church and the state were all looking pretty foolish. So, as we have discussed, he found a way to flip the script on personal assurance.

Remember that Descartes, frustrated by the fickle foibles of the personifying empathetic powers of Christendom (which most agreed in his day had been horrifically infected by the cares of this world), resorted to leaning hard on his own understanding. Whether it be true or false, he argued, he couldn't help having some form of understanding. At least he couldn't deny that some form of thinking was going on, because if he worried about whether he was actually thinking, that just made him think more!

So while it seemed Descartes had found a foundation for truth in his own first person perspective, his first act as a personally re-assured thinking thing was to turn to God and return the favor. From "I think, therefore I am," Descartes quickly jumped to "there must be a personal god who made me this way". And so, for a least a few hundred years after Descartes, the thinkers who followed him (we now know them as Modernists) believed first in the certainty of their own minds, and then secondly the minds of God and other people.

Let's pause here for a second. See what Modernism did to our essential selves? The individual, God and other people were all reckoned to be essentially thinking things. This meant that our lives– our souls– our spirits– the essential essences of who we are– basically just became our minds.

> *Our souls are not just our minds!*

But what about folks who weren't thinking things? The mentally ill or severely handicapped? Very young children? Deep sleepers? The very, very drunk? Well, let's just say they were all "problem cases".

But Modernist "insane asylums" were definitely not places you would want to live! You might even call them "soul crushing". It's no accident that one of the most powerful critiques of Modernism was Michel Foucoult's "Madness and Civilization: A History of Insanity in the Age of Reason", first published in 1961.

The point here is that Modernists were so fixated on the first person perspective– so excited about human minds– that they conflated the idea of a mind and a human soul. Many (but definitely not all) ancient Greek philosophers made this move as well; so Modernists didn't invent this problem necessarily, but they definitely threw some gasoline on the fire.

Breaking free from this Modernist mindset is one of the challenges we are attempting to address in this chapter. Our souls are not just our minds! But we will have more to say about that shortly.

Not too long after Descartes– let's say, by the late 1800's– belief in God began to lose some of its power. What if, some began to argue, it is our belief in the Judeo-Christian God that is holding us back from the elevated status we deserve?

Conveniently, there was a whole group of people who self-identified (and were generally acknowledged) as the people of the Judeo-Christian God. If we could successfully depersonalize them, this line of thinking went, we could justify ourselves as the ultimate authority over the most coveted power in all creation. We would be like God himself.

Especially since the world's most powerful people had already, for quite some time, been objectifying (ie. de-personifying) the lower classes and all people of all other civilizations (particularly Africans, but also Asians and all the "indigenous peoples" of lands they wanted to rule), this project of objectifying "the people of God" seemed particularly appealing.

If the personhood of the Jews could be denied, we could effectively prove our place in the universe. And not only that, we would no longer have to feel those twinges of guilt and self-doubt when we objectified other, lesser beings.

You might think God would have come down from heaven right then and there and publicly squashed the head of the snake that whispered those evil lies into the hearts of Modernist men. But He is more patient than that. He didn't crush the head of the serpent in the garden, and he refrained again when the serpent's lies gained traction in the early 1900's.

The Holocaust happened. And instead of the Judeo-Christian God stepping into human history to stop it, he did the same thing he did in the garden. He let things play out. After all, we wanted to know what evil was.

He warned us against eating such fruit, because he knew that in tasting it, we would taste death. The serpent and his lies will all be crushed eventually, but . . . we are getting ahead of ourselves.

In the early 1900's, European powers sought equality with God and came away with the stench and full, evil taste of death all about them. God didn't directly crush those forces of evil. Not yet. In the 1900's, like in the story of the Tower of Babel, it was squabbling among the human rulers themselves that brought the whole project down.

By the mid-1900's, they had fought to a standstill. Eventually, we learned that no mere human has the power to rule the world. The best we can hope for is to destroy it. And so world history entered into a new era. Some call it Postmodernism. Whatever it may be called, it is ruled by a global military doctrine that represents well the tragic absurdity of humanity's desperate attempt to hold on to power. It is known by the acronym MAD– Mutually Assured Destruction, and it has been the sole arbiter of world peace since the mid-to-late 1900's.

Over half of the world's population has now lived their entire lives knowing that one political miscalculation could end human civilization within minutes. Most of us would die without even knowing what was happening. The rest of us would freeze or starve in the consequent nuclear winter.

No one even bothers with nuclear fallout shelters anymore. They just aren't worth the effort.

And yet we keep living and working and empathizing with each other. We might as well be nice to each other, right? Thus the slow but steady rise of the power of liberal democracy so nicely summarized and wisely advocated by Canadian philosopher Charles Taylor (and others).

> *As long as none of us seeks exclusive control over the power to attribute personhood, we can all share it.*

The still-somewhat-new Postmodern peace agreement is roughly this: As long as none of us seeks exclusive control over the power to attribute personhood, we can all share it. There is no need for God, but the system only works as long as we keep working together, reassuring each other that we all have the power to define ourselves however we want.

But, as I mentioned earlier, this system of peace and mutual personification only works if it is totalitarian. Everyone must be allowed to define their own identities against the moral fabric of society. And this means everyone . . . without exceptions. All it would take is one slightly contagious, de-personifying belief in the wrong place to bring the whole world to an fiery end.

## 4. The Second Person Perspective and Christian Hope in the Face of Death

When Jesus sent his twelve Apostles out into the world, he warned them of persecution. But in Matthew 10:28, He told them not to fear those who "kill the body but cannot kill the soul."

Let's think about what that means.

As we mentioned earlier, throughout most of the last 2,000 years, Bible readers have brought a Greek, substance dualist worldview to the text. As a result, most readers have understood Jesus' words to mean something like, "If people kill your bodies, don't worry about it. Your souls are immortal. If they kill your body, you will just live on as a disembodied soul, and the life of a ghost really isn't all that bad. No more pain. No more suffering. It's really just the ultimate weight-loss program."

This reading, no doubt, has brought some comfort to many in the face of death. They believed the Bible was telling them that the death of their bodies was just another ordeal they had to go through– a painful step in their journey to a truer, fuller life in heaven as a disembodied, immortal soul. Hopefully they didn't consider Jesus' words too carefully though because his next sentence would have taken that comfort away!

"Rather," continues Jesus, "fear him who can destroy both soul and body in hell."

Wait. What? Bodies can go to hell? What about the idea that heaven and hell were places for disembodied souls? And . . . even worse . . . what is this business about the soul being destroyed? Aren't souls immortal? What is going on here? Jesus must not have known his Greek philosophy very well.

But if we reject our Greek, substance dualist presuppositions, what's left? How are we to understand Jesus' words? Even more importantly, what hope do we have in the face of death? And if a soul is neither a disembodied first person perspective (as we discussed in the last chapter) or a ghost, what is it?

First, let's deal with the hope question. That's definitely the most important. So let's be clear: Christianity does not teach us to hope in the immortality of our own souls. Far from it! Our hope, both in life and in death, is in our unity with Jesus.

> Jesus must not have known his Greek philosophy very well.

105

As Christians, we are not our own. We were bought with a price. "If we have been united with him in a death like his," Paul says in Romans 6, "we shall certainly be united with him in a resurrection like his."

The joy of Easter centers around the empty tomb. The resurrection of Jesus was a big deal. And it was physical. He wasn't just resurrected figuratively or metaphorically, "in the hearts of believers" or anything shallow and dismissive like that. He was really, literally, physically raised from the dead. There was literally an empty tomb. It changed history. That is how God demonstrated His power over death– not in disembodied, ghostly immortality but in actual, physical bodily resurrection.

I don't want to take too much away from our discussion of the resurrection of the body in our next chapter, but . . . for now, let's just say that any feeling of loss we may have when we discover that life as a disembodied soul is not what we have to look forward to when we face death is . . . well . . . let's just say it is worth it.

> *Our hope, both in life and in death, is in our unity with Jesus.*

So if it is neither a first person perspective nor a ghost, what is a soul?

In our historical introduction, we said that human souls are just human lives. That is probably the simplest way of defining the word (and differentiating between the old, substance dualist conception and a more up-to-date understanding), but compared to all the ways we often use the terms 'soul' and 'spirit', it is probably also the most mundane.

Our souls are our lives, true. But they are also our essential selves, the deepest and most lasting sense of who we are. To say it more precisely: our souls are the sum total of our lives from the second person perspective.

So let's differentiate our souls from our minds and our bodies:

Our minds are our selves, viewed against the backdrop of ideas. They are best known from the first person perspective. As such, our most trustworthy knowledge of minds is subjective. If someone says they are in pain or thinking about white elephants or that they believe in God, for example, it is impossible to prove otherwise. They may be confused about certain concepts (if they say they love pain, for example, it may turn out that they simply have the definitions of the words 'pain' and 'pleasure' mixed up) but their subjective experience is the sole arbiter of truth regarding their minds.

Our bodies are our selves, viewed against the backdrop of objects. They are best known from the third person perspective. As such, our most trustworthy knowledge of bodies is objective. If someone says they weigh 350 pounds, for example, it is very easy to prove whether or not what they are saying is true. We grab another object, a scale, and properly orient the two objects– we get them to stand on it. If the needle on the scale points to 250 instead of 350, we trust it's testimony more than the person's. Objective knowledge trumps subjective opinion when it comes to bodies.

Souls are trickier than minds and bodies. Our souls are our selves, viewed in relation to other persons. They are best known from the second person perspective. As such, our most trustworthy knowledge of souls is intersubjective. If someone says they are a pretty good guy, for example we don't just take their word for it. Likewise, there is no physical object– no "soul scale"-- we can use to measure a soul. If we want to know about a person's soul, we ask other people about that person.

But we can't just ask any other people. They have to be qualified.

In case that "subjective", "objective" and "intersubjective" language has you scratching your head a bit, let's think about a familiar, specific example: Adolph Hitler.

I am sure Hitler thought he was a pretty good guy. He certainly wasn't lazy, directionless, or sloppy when it came to his personal hygiene. But . . . he was definitely . . . not . . . a pretty good guy. His subjective experience would matter quite a lot if we were trying to figure out what he was thinking (whether he was in pain, or "really" planning to ally with Mexico, for example), but it really doesn't matter that much when we are trying to get to the truth about his soul.  The kind of knowledge we are looking for when we want to evaluate a soul isn't really subjective.  So what is it?

Maybe we could judge his soul objectively.  Ok, let's look around in the world of objects, find his soul and analyze it.  Is there such a thing as a soul scale?  No.  Unfortunately not.  But even if there was such an object as a soul scale, we'd have trouble finding a soul to measure with it in the world of objects.  That's just not the way a soul is.

A soul is not an object, so it can't be judged objectively.  So does that mean it can't really be judged with certainty?  Of course not.  We know better than that.  So there must be another kind of knowledge that is neither subjective nor objective.  That's why I was using the word 'intersubjective' earlier.

I didn't make up the word 'intersubjective', but it's not exactly standard vocab for small talk either– at least not where I'm from. But just because it's not common doesn't mean it's meaning is terribly complicated.  Whereas the word 'subjective' refers to your own, private thoughts, 'intersubjective' just means "what other people think".

> A soul is not an object, so it can't be judged objectively.

Our souls are best understood neither objectively nor subjectively.  They are most properly

109

understood intersubjectively. That's right. If you don't care about what other people think, you don't care about your own soul!

But it really isn't quite as simple as all that. Remember when we said, "If we want to know about a person's soul, we ask other people about that person– but we can't just ask any other people." Let's unpack that a little more.

Hitler's SS officers probably agreed with him on a lot of stuff. They probably even thought he was a pretty good guy. So does that give Hitler a pretty good soul? Not hardly! Their endorsement of his soul would look a lot like Wayne's idea of knighthood. It would have no real weight to it.

Taking this point a little further– if we were to call Hitler a saint, we couldn't be referring to sainthood in the normal, Christian church-defined sense. We could call him the patron saint of fools, or of racists, or of racial supremacists, for example, but that would be saying less about him and more about the kind of people who judge his soul positively. See how this works?

So if there really is a truth pertaining to the merits of his soul, where would we find it? Whose opinion would carry weight? Answer: the fool's opposite. The Bible says wise people are to judge the merits of souls (see 1 Corinthians 6). But ultimately, it is only the God who originally personified humans who can (and will!) ultimately judge our souls.

If this feels like a new and questionable way of thinking about our souls/spirits, go back through the

Bible and try it on. When 1 Samuel says the souls of David and Jonathan were knit together, for example . . . or when the Psalmist speaks of his soul being uplifted . . . or when death is described as a person's spirit leaving them . . . it is hard to find a Biblical example that makes more sense using the old substance dualist concepts than when we understand a soul to be a life viewed as a whole from the second person perspective.

This is especially true when we think about the souls awaiting judgment in Revelation 6:9-11. Do we really think the Bible is saying the minds or ghosts of the martyrs have been stuck there saying "How long, how long, how long" for thousands of years? Of course not! But neither are their lives forgotten. The testimony of their souls is ever before the throne of God.

We may also want to consider the ways we stretch spiritual language in ordinary usage to describe non-human things. When we talk about the soul of a university, for example, or team spirit, are we talking about the University or team's first person perspective? Their thoughts? Are we talking about any kind of ghosts? Of course not! We are simply talking about their quality and strength of vitality as if they were persons (ie. from a kind of second person perspective). We are

> *It is only the God who originally personified humans who can (and will!) ultimately judge our souls.*

111

referring to the ultimate, essential, or most fundamental unifying vitality of the university or team.  And it is both good and proper to talk about these intersubjectively understood things.

## 5. The Social Construction of the Soul

I tried to come up with a scarier title for this section, but this is the best I could do.

Socially constructed truth is one of the most hotly debated concepts of the postmodern world.  It wasn't very controversial in the pre-modern world and modern worlds simply because it was wielded exclusively by rulers.  But the rise of liberal democracy has led to increased squabbling over this sort of thing.

So let's think a little more about the intersubjectively understood kinds of things commonly referred to as "socially constructed truth".

> *It is easy to get confused between "objective truth" and "ultimate truth"*

Aside from our souls, money is probably the clearest and least controversial example, so let's focus on it for a bit.

Most people who think about these things guess it was pretty early on in the history of civilization that people began using trade tokens. They were just so much more convenient than simple bartering . . . for all sorts of reasons. Eventually gold became the most common trading token, but silver, gems, and even colored beads were also used from time to time.

As civilizations became more advanced, rulers began vouching for the authenticity of these trade tokens. They began having their faces and other sorts of things stamped on coins. They "unitized" the tokens, so that one token could be used to stand for lots more than one unit of money. They developed special formulas used for printing bits of paper representing different denominations of units and eventually they began . . . doing . . . whatever it is they now do to verify digital bank transfers. You get the picture.

But the point is that money (tokens, papers, or just verified digitally recorded numbers) has value simply because we all agree it has value. And its value (both the basic merit and extent of its value) is socially constructed.

This example gets even more exciting as we move into more modern history– as we have gotten even more comfortable with the sheer power of intersubjectivity.

In the United States, for example, paper currency used to be "backed up" by gold or silver. Each piece of paper currency printed was tied to a specific object stored somewhere on behalf of the person holding the paper. That way its value was considered "objective", see?

Lots of people felt good about the arrangement, but lots of other, smarter people thought it was silly. Why bother with the gold and silver? If we all agree that the paper is valuable, isn't that enough?

Well yes, actually.

As it turns out, our agreeing that dollars have value is, in fact, enough. In 1971, the US government announced that it would no longer try to hold on to any gold or silver corresponding to its dollars. And now, several decades later, dollars are the most trusted currency in all recorded history.

Boom.

But wait! Doesn't that mean that dollars no longer have any objective value? Technically, yes. But that's OK, because the value of money isn't grounded in the third person perspective anyway.

That's not how it works. Money only has value because we agree it has value. And it is only worth as much as we think it is. That is why central bankers spend so much time and energy trying to manage . . . well . . . you get the point.
Objective truth isn't the "gold standard" of truth after all. (Or maybe it is the gold standard, but we now know that the gold standard doesn't really matter all that much.)

Some Christians (especially Bible-believing Christians like me, who consider Fundamentalism part of their theological heritage) may scratch their heads at this. After all, objective truth has been pretty darn important to a lot of our people. But it shouldn't have been.

It is easy to get confused between "objective truth" and "ultimate truth", but it's also pretty easy to demonstrate the difference. After all, as Bible believers, we recognise that Jesus is the Truth (see John 1), and He is clearly a person (not just an object).

> *God keeps medium sized objects very consistent because he promised to.*

The way I understand it, ultimate truth (demonstrated most objectively in the person of Jesus Christ) is not merely an object. He is much more than that. Truth has existed from all eternity–well before any objects were created.

All mere objects were created by Him and for Him. Ultimate truth is revealed in creation, but not fully. There's lots of stuff no mere creature (or group of mere creatures) will ever know.

Nor is ultimate truth subject to any form of creation other than the Creator's own promises. He can make objects come and go as He pleases! The blind can be made to see. The lame can be made to walk. The dead can be made alive again. And we may never know how He does all that He can do!

God keeps medium sized objects very consistent because he promised to (see Genesis 8 & 9), but we would have to be absolutely, terribly, horribly mixed up to say that the authority of objective truth has more clout than He does! But I digress.

So, why is it that second person (intersubjective or socially constructed) truth is so controversial these days? That answer isn't nearly as complicated!

The problem with socially constructed truth is that we have to agree on it. And (in case you haven't seen the news since Walter Cronkite) there is quite a bit of disagreement in the world these days.

There is always ultimate truth (what God thinks), but we are not as able to "think His thoughts after Him" (as Johann Keper was fond of saying) regarding some subjects as we are with others. We do an OK job of agreeing when it comes to prices and things like that, but when we talk about other socially constructed truths like race and gender . . . let's just say I'm happy to stay in more reasonable waters. I am thankful for the clarity of the Bible when it comes to stuff like sexuality, but gender is another ball of wax. In Scotland, where I did the initial work for my PhD, they used to think running around in skirts was a manly thing to do!

And that is about as far as we need to go down that road. Let's try to stay on track.

For our present purposes, all I am trying to say is that our souls are socially constructed. When I talk about my soul, my "deepest core" or "most closely held values", I am talking about the things that make me uniquely me. My wife, my sons, my closest friends– they know my soul pretty well. And if I am maligned by the media or political enemies or mean book reviewers, I can trust them to resist the lies and hold on to the real me. But even if they forget me or disown me, God won't. And ultimately speaking, that is where my soul is most secure.

## 6. Reluctantly coining a new term– the PREsoul

There is such a big difference between the soul as it is known by God and the soul as it is known by mere human society that it would be difficult to overstate. Philosophers and theologians often use the phrase "infinite, qualitative distinction" in cases like these. So maybe we should reserve the word 'soul' to refer to our lives viewed as a whole by God and use another term to refer to our lives viewed as a whole by other persons. After all, even though both terms refer to human beings from the second person perspective, God's understanding of us is so much more important than everyone else's . . . Maybe there is already a better term we could use, but I can't think of it.

God's understanding of our lives, viewed as a whole, is what we most properly mean when we talk of 'human souls'. His perspective is the ultimate truth of the matter. But our reputations (ie. our selves defined in the moral fabric of society) are pretty similar and also super important. We could even say that they are our temporal society's best estimate of the souls God knows most authoritatively.

So let's take a lesson from academia.

Since human scholars have acknowledged for some time that there is a big, important difference between ultimate truth on the one hand and our best estimation of that truth on the other, they developed a system of verification known as the peer review process. If you are an expert in your field and you want to share something you think you know with other experts, great. Say it. But do the rest of us then just have to take your word for it? No. We certainly should sometimes, but . . . we all know that even experts make mistakes.

So what could carry even more weight than the word of an expert? Answer: the word of an expert that has been studied and deemed valuable by other qualified experts. This form of reasoning gave rise to the "gold standard" of academic publication. Before these kinds of publishers publish the words of an expert, they have other experts review it and vouch for its value. Pretty smart, right? This process has come to be known as the peer review process.

Similarly, there is a big, important difference between our souls-as-understood-by-God and our souls-as-understood-by-mere-humans. The former is the ultimate truth. The latter is simply a peer reviewed estimate. So until someone else comes up with a better term (or simply reminds me of a term that already exists), I will refer to this later concept as the "peer reviewed estimate of a soul, or a 'PREsoul'.

## 7. And in the End

Now, returning to the words of Jesus we discussed earlier . . . Do you know of a place in the Bible that says God knows you? Your true soul, not just your PREsoul? And not just human beings in general, but you in particular?

The words that come most readily to my mind are "the hairs of your head are all numbered". That's pretty specific knowledge, isn't it? And it wasn't just some random accounting office that numbered those hairs. These are the words of Jesus himself! Those are the words that give me comfort as I contemplate my own mortality– as I face the end of my life in this age.

The integrity of my life– my soul, my spirit, my deepest sense of self– will not fully depend on my friends and family or anyone else limited to human history remembering me when I die. When, as Ecclesiastes 12:7 states "the dust returns to the dust it was and the spirit returns to God who gave it," I will take comfort in Jesus's promise that He knows me in such detail. He will sustain my soul when I am dead. In fact . . . let's look a little more at the broader context of those words of Jesus we examined earlier:

> And do not fear those who kill the body but cannot kill the soul. Rather fear Him who can destroy both soul and body in hell. Are not two sparrows sold for a penny? And not one of them will fall to the ground apart from your Father. But even the hairs of your

head are all numbered. Fear not, therefore; you are of more value than many sparrows.

I know that I have more value than many sparrows to Jesus. Not just because He tells us so in this passage, but also because He showed us how much he values us by dying for us. And He didn't let the story end there! He died and was placed in the grave, where He remained for three days. But then God's power over death was demonstrated through Jesus' resurrection.

> *He knows my soul as it has been cleansed by the greatest sacrifice the world has ever known.*

I know that (unless Christ returns beforehand) I will die too, and be placed in some form of grave. I will remain there for an indefinite amount of time, but my story will not end there. I also know that, at a deep, spiritual level, I have been identified with Jesus. I have placed my hope in Him, and I have responded to His saving work with gratitude and acts of service. He loves me. He has called me into His service. He has saved and is saving me.

Even if I don't focus my consciousness as I should throughout life and/or in the face of death, my hope in the face of death isn't threatened. My hope is not in my first-person-perspective-self. It's not in my mind. Jesus never promised to make me super smart, or keep my consciousness going without ever stopping.

Even if I don't take care of my body as I should or it is ravaged by sickness, by injury and/or by slow surrender to the entropic forces of this world, my hope in the face of death is not threatened. My hope is not in my third-person-perspective-self. It's not in my body. Jesus never promised to keep my body alive without dying at some point before He returns.

Even if I annoy, offend and injure everyone around me, my reputation is shot, and the world hates me, my hope in the face of death is not threatened. My hope is not in my second-person-perspective-self-from-my-peers. It's not in my PREsoul. Jesus never promised to keep me from being misunderstood and hated by folks in this life– in fact, He directly warned us He wouldn't.

My hope is built on nothing less than Jesus' love and righteousness. Even in the face of the most extreme confusion, suffering, malalignment and death imaginable, He upholds and restores my soul. He knows, loves, preserves, and even refines my true life– my soul– in a more beautiful form than anyone other than He has ever known it.

He knows my soul as it has been cleansed by the greatest sacrifice the world has ever known. And when He returns and resurrects my frail, broken body, He will restore my soul like it has never been restored before. I will experience joy like I have never experienced joy before. And I will know Him and everyone else He has redeemed with more clarity, stronger love and deeper fellowship than I have ever known before.

> *If we are to live in the presence of a holy God for all eternity, evil itself cannot survive; only the memory of evil.*

Furthermore, I take comfort in the fact that every part of me that is not aligned with Jesus will not just taste death, but suffer death forever. And this doesn't just apply to me. Our global population, our nations, our churches, our families, and even our very selves before they are fully adopted as sons (cf. Romans 8:23) will be swallowed up in death. And they will be with us no more.

But He knows the "us" He has redeemed. And the parts of us that are united with him– every hair on our heads and every affection we fostered– will be again in the future.

And that is as it should be.

If we are to live in the presence of a holy God for all eternity, evil itself cannot survive; only the memory of evil. And then, when those parts of our souls that are aligned with Him in a death like His are resurrected in a resurrection like His . . . that is the kind of hope that can truly sustain us in the face of death!

We are dust with a role and a relationship. All the dust of this world (including the dust that is us!) is currently under a curse because of our sin. But it won't be that way forever. The curse has been broken and we now live in anticipation of a new heaven and a new earth— dust that has never and will never be cursed. What we will be at the resurrection can now only be known partly. Eventually we will know much more, and be much more fully known. And it will be awesome! Forever!

For now, let's turn to our next section, where we will talk a lot more about our bodies and the importance of the resurrection.

## Chapter 6: The Third Person Perspective

### 1. From "What?" and "When?" to "Where?"

Compared to the ground we have covered in the preceding sections, our discussion of the third person perspective should be fairly straightforward. From the third person perspective, we are just our bodies, right? Yes, but it's not quite that simple. We still have some important issues to discuss.

So far, we have argued that substance dualism– the old view that we are essentially souls or minds temporarily entangled with the stuff of this world as long as we live, but set free by death to live on as disembodied souls– is not what the Bible teaches. Instead, we are dust that has been separated from the rest of creation, given an important job to do, and empowered to do that job through an especially intimate relationship with our creator.

We have argued that our minds aren't really "things" at all. They are not substances. They are simply our selves (specific bits of creation) from the first person point of view. The experiences we have as minds– that constant stream of thoughts we call "consciousness"-- is just the first-person perspective of what it's like to be a reasonably healthy and normally functioning human.

Similarly, our souls are not really "things", or substances, either. They are simply our selves (specific bits of creation) from the second person point of view. We also acknowledged that different kinds of people can have different kinds of second person perspectives on our selves– and that God's perspective is the one that matters most: particularly as we face death (and eventually judgment).

Now, as we move into a more specific discussion of our selves from the third person perspective, it is time to think more about these specific bits of dust that we are.

To me, the trickiest bit of thinking we have done so far was working with the Apostle Paul's references to being absent from the body.  Those passages have had me scratching my head for a long time, and . . . I really think I have it right, but . . . it's hard because I have heard so many people I respect share their understanding of those verses through a dualist framework.  As we discussed, I think the key to understanding those passages is understanding that when Paul is contrasting life "in the body" with life "with Christ", he is making a temporal contrast, not a spatial one– not "where did he go?", but "when did he go" (Cue the Tardis sound effect, right?).

And I really think that is the right way to read those passages.  But what about Paul's corpse?

After all, when people die, it makes sense to say that "they"-- or at least their first person perspectives– jump into the future, but . . . the Tardis (and contents) always just sort of disappears when it leaves a timeline.  In real life, a corpse is left behind.

That is the problem we will be dealing with in this chapter. We have talked about *what* we are (dust with a role and a relationship), and *when* we are (after creation, fall and redemption but before glorification). Now it is time to talk about *where* we are (specifically as we relate to physical bodies).

## 2. Where is your mind (and where does it go when you die)?

One of my favorite memories of Uncle Sam (who is, unfortunately, resting in peace now, having died a few years ago) is hearing him tease my middle son, Asher.

Asher was around 5 years old, and Uncle Sam was teasing him about girls. "You gotta be careful around girls, Ash." Sam said. "They will make you lose your mind!"

Asher looked back at him with an adorable expression of shock and alarm. "Wait!", he exclaimed. "I don't think I even know where my mind is!"

I doubt we can blame girls for Asher's problem. It probably had more to do with being raised by an overly philosophical dad.

Now it's time to address that issue . . . head on.

We all have a pretty firmly established notion of our minds depending on our brains. After all, brain damage affects our first person perspective pretty radically. A simple bump on the head can cause us

127

> Where do our minds go when we "lose consciousness"?

to lose consciousness. But it's not just our brains. Even stubbing your toe can interrupt your train of thought. Getting drunk or spinning in circles affects our minds in pretty predictable ways as well. So clearly our minds are connected to our bodies– more than just our brains. But what part of our bodies? Where in your body is your mind, exactly? Careful! Don't slip into the old dualistic trap again!

The dualist would have us looking around in our bodies for something that looks like a mind. Maybe our heads could be opened up to find a slimy little alien-looking thing pulling on levers and watching little computer monitors. Is that kind of what our brains are? Little hunks of mind-stuff that we carry around in our skulls?

We know that's absurd.

Maybe our minds are just really, really small– so small they are "non-physical". Maybe they are like, the size of mathematical points– that move around in our heads and sometimes, lightning quick, down to our toe when it gets stubbed. What do you think?

I think that would just shrink a big, really silly theory down to a really small, but still really silly theory.

In 1949, Gilbert Ryle published a book called "The Concept of Mind". In it, he criticized substance dualism as "the dogma of the ghost in the machine." What a great line! He demonstrated the silliness of that kind of thinking so clearly that substance dualism was pretty much stamped out of the academic discipline of Philosophy. The only problems were, 1) he didn't present a very convincing explanation of what a mind actually is (only what it isn't) and 2) when substance dualism retreated from the academic discipline of Philosophy, it barricaded itself in the academic discipline of Theology, which was already a mess for other reasons.

If not for these problems, I probably wouldn't have spent so much time worrying about it, and you wouldn't have spent so much time reading this book. But the entire Western world was (and arguably still is) reeling from the trauma of the world wars, so we can hardly blame Ryle. "The Concept of Mind" is a massively important book, even if its legacy has (so far) been more destructive than constructive.

Ryle argued that looking for the ghost in the machine (or a mind in a body) was committing a category mistake. It's like taking two apples, putting them together and looking for the number 'two'. Are there actually three things? Two apples and one number (the number 'two')? What about the color? With two apples, you've got some color, too right? And taste and texture and weight . . . Wow! How many things can you juggle?

No. Of course not. That's just silly. Numbers aren't *things* like apples are. Neither are taste and texture and any number of other properties. Does that mean numbers aren't real? No. Of course not! They just aren't substances. They pertain to stuff, but they aren't stuff in and of themselves. Same with minds. They aren't *in* normally functioning human bodies, they are *what it's like to be* a normally functioning human body.

Simple, right? Well . . . maybe as long as the human body is functioning normally. But what about when our bodies aren't functioning normally? Like when we are "put to sleep" for major surgery? Or when we die? Where do our minds go when we "lose consciousness"? Can we just say we lost it for a while and leave it at that? If so, how do we know that the mind we "lose" when we lose consciousness is the same one we find again when we "wake up"?

That might not seem like much of a problem at first, but if you let yourself think about it . . .

A friend (Holly) told me recently that she was afraid to go to sleep as a middle schooler. How could she know that she and the girl who woke up in the morning would be the same person? Falling asleep might actually be the end of her. Eventually she would get too tired and doze off. Then she would wake up in the morning, relieved to still be herself. But then 18 hours or so later . . . same problem.

Now you might be tempted to call me (and middle-school-Holly) silly. "What do you mean, how do we know it's the same mind?" you might ask. "We just know."

That's true. We do "just know." But, just for the sake of argument, let's play this out. If someone woke up from surgery with someone else's mind (I am sure there is a movie out there somewhere like this!) friends and family would immediately know the difference, right?

But what if the patient didn't give it away immediately with a comment like "Whoa, what am I doing over here? That's my body, over there!". What if, for example, a patient wanted to switch bodies and the whole thing was part of some evil scheme?

> *It isn't just our bodies that make us who we are.*

To be clear, this would be absolutely, positively, physically impossible for a long list of reasons, but the most important one for our purposes is that, even if they could somehow manage the mechanics, they wouldn't get away with it for long. Their closest friends and family would know the difference. The memories, values, habits, virtues, talents, style and taste– they might try to act like the other person, but they'd never get away with it.

We are, in an important sense, our bodies. That is why any kind of "mind switch" would be so impossible. Our minds are the first person perspective of particular bits of dust– our bodies, and literally every time we think a thought, there is a change in our bodies that corresponds to that thought.

But it isn't just our bodies that make us who we are.  There is a lot more that goes into our identities than physical or mental continuity.  Even our intersubjectively known PREsouls are just estimates of our true selves.  There is a big difference between subjective, objective and intersubjective (ie. mental, physical, or PREspiritual) truth on the one hand and ultimate truth on the other, remember?  Ultimate truth is a special kind of second person perspective truth.  It is what God thinks.

We are our bodies.  And we are our minds.  But, as we discussed in much more detail in the last chapter, from an ultimate perspective, it is our souls that make us who we are.  And when we talk about our souls, we are not just talking about our PREsouls (although our personal reputations are probably under-valued these days).  Our PREsouls are merely the "peer reviewed estimate" of our true souls– our lives from the ultimate, second perspective.  Ultimately speaking, however, the value of our PREsouls pales in comparison to what our creator, sustainer and redeemer thinks of us.  It is the second person perspective (particularly the "I-Thou" relationship of being known by God) that carries the real weight of ultimate truth here.  We will talk about that more in the next section.

But for now . . . what about the corpse?  As we have seen, the mind is what it's like to be a normally functioning human body.  While I live, the best answer to the question, "David, where is your mind?" is "I'm right here, see?  I'm awake and listening!"  But what about when I am not awake and listening . . . and won't be again until the resurrection? What will my body be then?

Sadly, a dead body is a body that no longer has a first person perspective. Upon death, it ceases to be a person. It becomes a memorial. It should be treated with dignity and deference because of who it was, and who it will be again in the future. But it is no longer a "who", but a "what".

A corpse is still human, but because its mind– its first person perspective– has shifted to the future, so does our ability to properly regard it from the second person perspective. We should treat a dead human body as an 'it', even as we hope and expect it to become a "you" again at the resurrection. And not just any "you", but the same "you" it was– only without the bits that weren't fully united with Christ.

## 3. Where is your soul (and where does it go when you die)?

Like the mind, the soul is easy to locate while we are alive. Where is my life? Where is my soul? While I am alive, it is right here, being lived by my body.

There are ways our souls can get entangled with other people, places and things, and I definitely don't want to discount that. The concept of "sentimental value", for example, should be very familiar.

It may be that your life is entangled with a letter or gold watch that your grandfather gave you. Or it could be a special place you go to be alone, to read, to pray or to cook dinner for your family.

Your life may become entangled with the lives of others around you, so that graduating from a school, losing a job or leaving a church can feel like a rending of your soul.

And then there is the sense of having our souls be elsewhere– like when the Apostle Paul said he was with the Collosians in spirit. Parents know that same feeling when their children grow up and start driving or (even more severe!) go off to college or move away for some other reason.

> *Our souls are designed for relationships.*

We acknowledge this entanglement of souls when we respect the sentimental connections of others. I remember, for example, when we discovered that our three year old son had left his favorite stuffed animal at a stop on a family camping trip. We backtracked several hours to search for it, and when we couldn't find it, the entire family grieved for and with him. I am still a little emotional even now, recounting the memory. We still have the favorite stuffed animals of our other two boys, safely stored in our attic, but there will always be one that is missing. Dang, that's sad! Why? Because our souls are designed for relationships. They easily become entangled, and this entanglement must be acknowledged for others to truly know us.

There is much more we could say on this topic of entanglement, but for now we should simply acknowledge that while we live our lives, the second person perspectives attributed to us are most properly directed at our living bodies. Just like my mind, while I am alive my soul is "right here". This acknowledgement sets us up for a more difficult question: where are you– where is your soul (the second person perspective on the person that is you) between death and resurrection?

Short answer: The same place you were between the time God first knew you and the time you were born– in God's plan.

Let's unpack that a bit.

We know that the mind (or first person perspective) jumps immediately from death to resurrection. From the first person perspective, when I die, I will lose consciousness and (unless I have some weird sort of medium like the Witch of Endor bring me back for a chat) I will "wake up" or regain my consciousness in the future, at the resurrection.

Just like your mind is the first person perspective of a special bit of dust, your soul is that same, special bit of dust from the second person perspective.

While you are alive and well, your soul is who other folks understand you to be. And remember that, just as different folks will have different understandings of you, not all of their estimations matter equally. In particular, we need to distinguish between your soul as God knows you (which, if you are a Christian, is like the ultimate glamor shot because of what Jesus did for you) and your place in the moral fabric of civil society (which we were calling your PREsoul in the last chapter, and can feel a lot like Wayne's version of knighthood compared to your soul as known by God).

You are a special bit of dust. You are a bit of dust with opinions and memories and values. You (as a special bit of dust) have morals and affections, and others regard you morally and affectionately.

At the resurrection, we will value you as the same, special bit of dust you are now even though the actual molecules of the dust will be different and you'll have none of your sins or sinfulness any longer. We will rightly regard your resurrected body as having the same soul it did before you died.

So where is the soul between the death and resurrection of a human person? What does death do to us from the second person perspective? I think Eccelesiasties says it best when it describes death as when "the dust returns to the earth it was, and the spirit returns to God who gave it."

There is a sense in which the souls (at least the PREsouls) of the dead are still around. We still remember them. We can learn things from them (cf. Hebrews 11&12– especially 12:1). We still know them and miss them and honor them. But they are dead. They don't have first person perspectives any longer. They can't see you. You can pray for them, but not to them. Praying for them isn't going to change them, but it might be good for you.

Similarly, it isn't proper to regard their dead bodies as people any longer. Viewing the body before burial or looking at old pictures and videos can be very good for the living, but only as long as it helps us understand that the relationship has changed– and look forward to the relationship changing again (beautifully!!) at the resurrection. So their souls are here with us, in the land of the living, but not in the same sense. We know their lives, but they are no longer living.

The lives of my great, great grandparents are important to me. So are the lives of my great, great grandchildren. But I don't know either one of them very well. I only know them as concepts. Thankfully, God knows them much better– but we'll get there in a sec.

> At the resurrection, we will value you as the same, special bit of dust you are now even though the actual molecules of the dust will be different and you'll have none of your sins or sinfulness any longer.

First, let's consider a somewhat strange but very helpful passage of scripture. In Hebrews 7:9 & 10 we read, "One might even say that Levi himself, who receives tithes, paid tithes through Abraham, for he was still in the loins of his ancestor when Melchizedek met him."

How can we say Levi, decades before his own birth, was with Abraham (his great granddad)? It certainly wasn't Levi's first person perspective self (or his mind) that was in the loins of Abraham. That would be . . . unthinkable. It also wasn't his third person perspective self (or his body). That's . . . just not how procreation works. It was Levi's second person perspective self. It was his soul that was there, in his ancestor, Abraham. Not his ghost, but his soul.

This simply means the identity of Levi is tied closely to the identity of Abraham. Why are Levi and his descendants (the Levites, or priests of Israel) important? You can't understand their importance without understanding that they are part of Abraham's legacy. "Father Abraham had many sons. I am one of them . . . so let's just praise the Lord!" If we are in Christ (our souls, that is,– see how this works?), we paid tithes through Abraham too. And we received due punishment for our sins on the cross. And we have been raised with Jesus, even as we await the fullness of that process– the future resurrection and/or transformation of our own bodies.

> *Neither our bodies nor our minds go to heaven. They never have, and they probably never will.*

In the same way, when God's people die we can say they– not their minds, not their bodies, but their souls– go to "the bosom of Abraham" their father. We aren't talking about an object (physical or non-physical) going to a place (this dimension or some other), but a concept moving from one class of concepts (lives being lived at a certain time) to another class of concepts (the lives of people not living at a certain time).

Similarly, we might refer to Christans now dead as "the church triumphant" because they no longer have to struggle against sin within and around them. On the other hand, those of us still in the struggle can be referred to as "the church militant". One group is made up of lives (or souls) now alive and active, while the other group is made up of lives (or souls) that are now inactive, or at rest.

The dead in Christ are not sleeping. They are dead. But the expression "rest in peace" is a good way of expressing the fact that, while they are no longer living their lives, they will live again at the resurrection.

So the PREsouls of the dead are still around (just not as still-living people) after they die. But thankfully, there is much more to the story! God remembers the dead much better than other people do. And with much more detail, authority and (for His people) grace. So, although it is important to remember that the dead do not have first person perspectives (or minds), in the most ultimate sense, the souls of the dead are with God in heaven. It is in this sense (and only this sense) that it is proper to say that we go to heaven when we die.

Neither our bodies nor our minds go to heaven. They never have, and they probably never will (unless we are treated to the kind of experience the

Apostle John had that produced the book of Revelation).

Our minds, or first-person perspectives, began on earth some time after we were conceived. When we die, we will lose consciousness until the resurrection. At the resurrection, our minds will still be the first perspectives of bits of dust. They are, and always will be earthly, whether we are talking about this old, cursed earth or the new, uncursed one that will be.

Similarly, our bodies begin on earth when we are conceived. When we die, they go to the grave as corpses, where they will be subject to decay until the resurrection.

But our lives (or souls), which have always been known by God, will no longer be lived by our bodies here on earth when we die. They will be again, but not between death and resurrection. During that time, sometimes referred to as "the intermediate state", they exist most explicitly in heaven. That is why we say that our souls, upon death, will return to God who gave them. They will remain there (in heaven), where they will be known by God in absolute, authoritative detail, until the resurrection. At the resurrection, our souls will be reunited with our bodies, which is simply to say that our bodies will begin living our lives again. The lives we live now will be the same lives we will live then . . . only much, much better!

## 4. Where is your body (and where does it go when you die)?

And so . . . what about the corpse? If it is not obvious by now, this is the question I've been avoiding. I'm not sure exactly why, but . . I guess the terrible, ugly tragedy of death affects all of us in different ways. For some reason or another, this is the part of human identity in the face of death that is the most difficult for me to process.

But I'll try.

At one point, I had envisioned this section as a narrative. I thought I would pick up the car crash story from the first person perspective section, but tell the story from a third person point of view. Instead of me crashing into an unknown, immovable object, it was someone I cared about.

"Dr. Norman, I am sorry to inform you that there has been a terrible accident. We believe your wife has died, and need you to come with us to identify the body."

I could imagine the state trooper there, standing at my front door with his hat in his hand. But when it came to actually going with him . . . seeing the body . . . nope. I can't do it. Not voluntarily.

I tried other ways of framing the thought experiment. Maybe I could imagine the victim was completely fictitious– a daughter even though in real life I only have sons, a sister even though in real life I only have a brother . . . maybe just a close friend. No. I can't do it. Not just because it is emotionally difficult. It just seems so . . . disrespectful.

So we are going to have to work through this section another way. But before doing so, please let me say that if you are dealing with the real life loss of a loved one right now, I am so, so sorry. I have tried, but I refuse to imagine your pain in great detail. I know the gritty, physical reality you are walking through right now is difficult. It is also deeply intimate, to the point of feeling sacred.

God has specifically called and gifted people to help you, and He will give you the help you need. It may be an especially kind or respectful state trooper on your doorstep. It may be the way a doctor or nurse or family member treats you. It may be the exceptionally, unexpectedly knowledgeable, personal, professional or understanding funeral home representative. It may even be that God gives you a supernatural ability to show grace to people who don't understand.

I don't know exactly how He will do it . . . but that is just the thing. God knows us and loves us so intimately that He regularly surprises us in the ways He cares for us and provides what we need– not always what we want, but what we would want if we knew what He knows.

Look for signs of His love and provision.

Remember that it's precisely the intimacy of that relationship that makes you different from all the other lumps of creation out there. That's the relationship that empowers you to do what you've been called to do.

Rather than try to work through anything as intimate as a narrative, I will simply lay out a few principles that follow from the understanding of human identity we have been advocating in this work. I will try to make them as practical, applicable, and concise as possible.

### 1. A corpse is a memorial.

When you encounter a corpse, it will command your attention. It is no ordinary object. It should be treated with dignity and deference, and you will know at a deep, visceral level that this is so.

But, sadly, a corpse no longer has a first person perspective. It is no longer proper (strictly speaking) to refer to it in the second person.

You may be tempted to address it as a person or try to talk to the person who died like you would pray to God (who does not have a body other than the now-physically-absent body of Jesus and yet still hears our prayers). You may even find this psychologically helpful. But, as may be obvious from our previous discussion, the person the corpse used to be will not be able to hear you.

The eardrums may be there, but the synapses in the corpse's brain are no longer firing. The person who was and will be again is not now. That's why it is appropriate to miss people who have died. The person is not "there" at that "point in time". The memories and the memorial may be there, but not the person.

At the same time, experiencing the sheer presence of a corpse can be a powerful testimony to the importance of the person's life, the tragedy of death itself, and the fact that the person is, indeed, dead.

Accepting the death of a loved one can be very important– especially when feeling the loss later or being confused by feelings of emptiness or neglect. If you don't process the reality of death fully enough, you may feel wrong for not relating to your loved one as a living person anymore. You may feel like a bad friend for not checking up on them or serving them in some way– like there is something you should be doing to continue the personal relationship.

Understanding the corpse as a memorial will help you relate to your loved one properly– as a person who was, is not now, but will be again in the future.

### 2. Respecting the corpse (as a memorial) is important. Preserving it is not.

God is good. When we sin, we demonstrate our desire to know evil. That's where death came from.

But that's not the end of the story. Not even close! It's just the beginning.

Jesus has defeated death and invited us to join Him on the winning side. He has opened the path to the tree of life. If we are committed Christians, He has given us eternal life. We have His Spirit active in us now– that's how we are able to resist sin sometimes– and we will be fully adopted as full recipients of eternal life at the resurrection.

But none of this is up to us. Death and decay are the result of sin. Jesus and only Jesus can undo the effects of sin.

We are not saved by our works, but by His work. Our works can't save us or preserve those we love. That's not the way it works. We trust the power of God (and only God) to save us (mind, soul, and body) from sin and death.

3. **Real and important decisions must be made . . . by real and important people.**

Decisions made by, around, and for a dying human being are some of the most grueling decisions ever made.

And they do not end in death.

If you are currently in the position of having to decide what to do with a corpse, sorry about the double-negative here but . . . please do not undervalue yourself.

Lots of people will have opinions about how the corpse should be settled between now and the resurrection. Some will share those opinions. Some will not. Some of those opinions will matter. Some will not.

Embalming, cremation, memorials, cemetery plots . . . even what to do with highly personal items ranging from false teeth to wedding bands– there are any number of decisions that must be made. Please, please, please do all you can to respect the person who must make these decisions . . . even if that person is you.

There is no mistake that can be made– by you or by anyone else– that will upset God's plans for that very special bit of dust.

### 4. A human corpse will not remain a corpse.

Whether carefully preserved or respectfully cremated on the one hand or neglected, desecrated, dissected or even fed to cannibals on the other, no human corpse will remain a corpse forever.  When Jesus Christ returns in glory, according to the Bible, every human corpse who has ever lived will be restored and re enlivened for judgment.

Those who are united to Jesus will be judged according to His merit.  Those who trust in their own merit or the merit of some other person or philosophy will also be judged accordingly.

Will that universal resurrection be hard for God? No. Nothing, of course, is hard for God. But some things require a miracle. I believe my conversion to Christianity required a miracle. So will my resurrection. God doesn't need my help, but He asks me to give Him my trust. But that's just to make things easier for me. Not Him.

## 5. Things could get awkward . . .

For those who are in Christ, resurrection day will be an awesome, happy day. But for those who have trusted in something else– their own merit or other false gods . . . not so much.

I am honestly not sure how much it will matter, but . . . it seems to me that resurrection day will be a happier day in some neighborhoods than others. That is why I want my remains and those of my loved ones to be buried in a church cemetery. I am somewhat saddened by the fact that our current local church doesn't even have a cemetery . . . but I don't think it will matter very much.

What does matter, on the other hand, is that we remember to actually believe in the literal, physical resurrection of human remains.

> *God doesn't need my help, but He asks me to give Him my trust.*

147

# 5. So . . . What's Our Hope in the Face of Death?

The subtitle of this book is "Personal Identity and Christian Hope in the Face of Death." We have talked a lot about personal identity, but when we have talked about death things have seemed kinda sad and dark, haven't they?

When death takes place slowly, having things finally end can be a huge relief. But is this relief the be-all and the end-all of Christian hope in the face of death?

Absolutely not!

The most eloquent summary of Christian hope I have ever seen is the start of the Heidelberg Catechism:

> What is your only comfort in life and in death?
>
> That I am not my own, but belong with body and soul, both in life and in death, to my faithful savior Jesus Christ. He has fully paid for all my sins with his precious blood, and has set me free from all the power of the devil. He also preserves me in such a way that without the will of my heavenly Father not a hair can fall from my head; indeed, all things must work together for my salvation. Therefore, by his Holy Spirit he also assures me of eternal life and makes

me heartily willing and ready from now on to live for him.

We are not our own. We do not define ourselves, nor do we set our own persistence conditions. Likewise, though we are dust and to dust we will return, we are not defined by creation. We are defined by the Creator.

The empty tomb of Jesus was a big deal. It was literal and it was physical. The bit of dust that hung on the cross until it stopped functioning as a living human being is the same lump of dust that was laid in the tomb, was fixed and put back into motion, appeared to the disciples and then ascended into the sky.

> *The persistence conditions of a person are set by the person bestowing personhood.*

But does that mean that every atom that made up Jesus' body on the cross had to be included in the lump of dust that walked out of that tomb? Every hair from his head? Every drop of his blood? Of course not!

We are dust, but we are not defined by dust. The molecules that make up you and me right now are not the exact same molecules that made up you and me a few years ago. And they won't be the exact same molecules that make us up at the resurrection.

So how many molecules can change before we have to call it a different body? Don't get sucked in by the paradox! Graciously, God closed off the path to that trap by creating us out of matter that is in a constant state of change. But that doesn't mean there aren't lots of smart people who have gotten tripped up by these kinds of details– so be careful. And the best form of care I know of is to remember what made the body such a special lump to start with.

My body is important to me. The bodies of my friends and family are important to me. But I'm not the one who made them important to start with, and I'm not the one who will make them important– and, importantly– them at the resurrection.

The persistence conditions of a person are set by the person bestowing personhood.

And there is only one Person who can set those conditions ultimately and for all persons. That's the Person Christians put their faith in when facing death. That's the Person Christians put their hope in when facing death. And that's the Person Christians love the most when facing death.

After all, there is no greater love than to lay down one's life for a friend. And we love because He first loved us.

# Part 3: Epilogue

Thanks for thinking this through with me. I know you won't have agreed with everything I have written. I may not in a few years either. So I want to underscore that the book you are holding in your hands is a first edition.

I expect to make edits. That is why I haven't ceded publishing rights to anyone. I have reserved the right to edit at any point or pull this work off the market completely and write a public apology. It may be your feedback that convinces me to do so. If so, I will thank you. Faithful are the wounds of a friend.

There are tons of details I have either glossed over or avoided entirely. That doesn't mean I have never thought about them. It just means I either forgot or didn't think they were of central importance to the main objective. You may convince me otherwise, in which case I will expand and/or add to this work in future editions. Alternatively, I may need to write more scholarly work to address finer points that turn out to be important to avoid misleading myself or others. I may not want to necessarily, but I am willing to do that work as well. Or . . . and here is a much better idea . . . you could do the scholarly work, and I will do all I can to help you and cheer you on.

My hope for this work is that it reframes a discussion that I believe is very, very important–

perhaps more important now than it has been in over 1,000 years. Will you participate in this discussion with me? Will you invite others in?

Let's clarify and share our understanding of human identity and Christian hope in the face of death. It is desperately needed by each of us, and especially the world that is lost and dying all around us.

Let's keep working.


David.

Made in the USA
Middletown, DE
06 January 2025